商用英文書信

篠田義明 監修

高崎榮一郎

Paul Bissonnette

著

彭士晃 譯

三民書局

監修者的話

　　書信的寫作，要使收信者不會產生誤會，而且能夠達到發信者預期的效果，實在不是件簡單的事情。本書就要挑戰這個目標，以實例來介紹實務現場圓融的溝通技術。本書的特色，在於全由工作場合中收集而來的實際範例所組成，並且加上註解。對在這個領域中工作的人士，或是即將出社會工作的年輕學子們而言，本書將是極佳的指引導讀。

　　這一類的實用書籍，往往流於指正英文句型結構或文法錯誤；但是，本書除了能解決這些問題之外，還追求一份文件在內容上應有的形態，亦即內容的邏輯結構。僅就這一點而言，應該沒有其它類似的書籍能與之媲美。我不斷地提倡，要達成良好的溝通，就必須要有明確的邏輯結構。這本書完全符合我的主張。

　　身為本書的監修，在初稿完成時，除了「本書用法」和「參考文獻」之外，其它的部分均詳讀過。對於作者引用在下的拙著，或是與我演講中主張的內容極為酷似的地方，均寬大待之；而對於其它我覺得不妥之處，也不客氣地加入自我的意見，並請執筆者修正。在這樣的作業下完成的本書，若能對各位從事實務的人士或莘莘學子有所幫助，才是我最大的喜悅。

　　作者之一的高崎榮一郎，是我 10 年前於東洋工程株式會社擔任員工技術書信教育講師時指導過的學生，他是位認真而優秀的員工。之後，當我擔任日本技術溝通協會會長時，他也在協會中致力鑽研。至於 Paul Bissonnette 先生原本是東京外語中心的英文老師，當他參加過我的研討會之後，對這個領域產生極高的興趣，並在美國獲得碩士學位，學習極為認真努力。希望他們二位今後能夠繼續不斷地鑽研。

　　最後，我要感謝研究社久田正晴先生的盛情與指導，使得這本書有機會問世，在此表達由衷的謝意。

1995 年 5 月 1 日
篠田義明

本書用法

本書中所舉出的範例原文，是由國人書寫的。經過 Paul Bissonnette 修改過的原文改善範例，會加上（○）的記號。此外，其它歐美人所寫的英文中，如果有可作為範例的，也同樣加上（○）的記號。而國人所寫的英文，經過以英文為母語者只針對文法或語法上的錯誤略作修改者，會加上（△）的記號，另外再附上 Paul Bissonnette 修改過的範文。

本書各節的結構大致如下：

(1)標題是將範例在溝通上的問題點作成簡短的提示，在標題之後，再針對本章所要提出的問題主旨加以說明。

(2)【背景】說明該範例寫作的經過與原因背景。

(3)接下來是「範例（原文）」，以雙細線方框框起。

(4)在範例的方框之下，說明範例中使用的主要語句。但有時為了避免重複，往往將語句的說明改列於後述的「改善範例」之後。

(5)【內容】介紹範例的大意。

(6)【問題點】詳細說明範例原文的問題點。

(7)之後介紹原文的「改善範例」。「改善範例」會加上（○）的記號，並以粗細線方框框起。

(8)在「改善範例」的方框之下，說明改善範例所使用的主要語句。

(9)【段落大綱】簡略說明改善範例的段落結構，希望能提供讀者安排書信架構順序時的參考。改善範例例文所標示的①、②……等號碼，代表歸納出來的架構段落。

(10)之後是改善範例的【內容】，說明改善範例的大意。

(11)【注意事項】說明改善範例與原文的差異。

⑿ （參考） 研討該節中所使用的溝通技巧和語法，給予適切的說明。

我們並未完全指出原文範例中所有的問題點和文法錯誤，因此，即使讀者已將【問題點】中所提出的問題全列入考慮，我們仍舊建議您不要拿這些原文作為參考。您可以參考的部分，是以粗細線方框框起，並附上（○）記號的改善範例。

本書中所使用的範例，並非刻意收集最糟糕的書信，而是作者在各個企業開設的內部研討會或日本技術溝通協會 (JATEC) 每月例行研究會中，與會諸君以匿名方式寫作的書信。使用這些書信，單純只是為了討論技術上的問題，絕無其它用意。由於出處並非特定，所以無法一一向原作者請求允許刊登；但也由於有他們的協助，本書才能順利完成，特此致謝。

由於 JATEC 成員在每月例行的研究會中努力學習，才使本書得以整理匯集成書。此外，JATEC 會長、同時也是早稻田大學教授的篠田義明先生，在本書寫作的過程中，自始至終惠予指導，並在百忙之中負責監修和惠予指正，在此表達最深的謝意。

1995 年 5 月 1 日
Paul Bissonnette
高崎榮一郎

商用英文書信

目　　次

監修者的話
本書用法

第III章 通知與回覆

第Ⅴ章 催 促

1. 尊重對等主義 *124*

2. 說明理由 *130*

3. 自己人也得注意禮節 *134*

4. 催促也得公事公辦 *137*

第VI章　附　件

第VII章　致　意

1. 問候的語氣應考慮對方　148

2. 與其回顧過去，不如展望未來　151

第VIII章　提　案

1. 利用洗鍊的英文範本　158

2. 建議、請求或要求　168

3. 從整體到細節　172

第IX章　簡　　介

第 *1* 章

交涉

1. 考慮對方的立場

在我國，中小企業向大企業低頭、賣方向買方低頭是理所當然的。這種習慣似乎已經根深蒂固，並隨著拓展世界貿易的腳步帶到海外，尤其是我國經濟在海外獲致成功，甚至可以得見這股想將自己的習慣推展給外國人的趨勢。

縱使有些外國企業積極地引進我國的習慣，但是我們也應了解有些人對於我國的做法是相當反感的。我們更不可以忘記，在歐美人的主觀意識中，總認為我們是只要歐美有好東西，就會馬上模仿的國家。

下面這篇例文，就是強要美國的小企業接受我們的商業習慣和感覺；但是這種商業習性，恐怕會引起外國人的反感。

【背景】

ABC 公司的田先生將美國小企業 XYZ 公司的 "New System" 介紹給日本的 DEF 公司，希望能將此一系統賣給該公司。但是，DEF 公司的金先生無法單憑 XYZ 公司提供的資料就決定該不該購買這項技術，所以提議希望能和 XYZ 公司的克拉克先生見面聊聊。至於克拉克先生雖曾經答應過與金先生見面，但後來希望在見面之前應先簽下保密協議。田先生認為要立刻簽下保密協議，恐怕會帶給金先生負面印象，所以寫了一份傳真給克拉克先生，希望在簽訂協議之前，再多提供一些資訊。

例文 1　促請對方積極拓展國內市場的傳真（原文）

I have heard from our New York representative that you want to postpone the meeting with Mr. King of DEF on September 15, unless he will conclude a secrecy a greement[1] with you.

If we want to sell new technology like "New System" to potential Japanese clients[2], in any case some preliminary (non-confidential base[3]) presentation should be absolutely necessary at least.　Otherwise they could not make their judgment even on conclusion of secrecy agreement.　Other sellers of same kind of system like yours are approaching to potential clients to present positively[4] the outline including the drawings of their system on non-confidential basis. Therefore, DEF, for example, has to know at first at least the outline and features[5] to be different from other systems already known to DEF before entering into[6] secrecy agreement.　I would ask you to reconsider to take a positive action.

I look forward to receiving your favorable reply[7] by return.

1) to conclude a secrecy agreement　簽訂保密協議　2) potential client 潛在顧客　3) non-confidential base　非以保密為基礎（= non-confidential basis）　4) to present positively　積極地提供　5) features 特色　6) to enter into ～　締結（協議）　7) favorable reply　善意的回應

【內容】

　　我從本公司紐約代表處聽說，如果不與您簽訂保密協議，您希望延後與 DEF 公司金先生在 9 月 15 日的協商時間。

　　如果想將 "New System" 這類的新技術銷售給潛在的日本顧客，事前提供部分資訊（不以機密的方式處理）是絕對必要的。如果不這麼做，顧客如何判斷應否簽下保密協議呢？其它與貴公司銷售類似系統的公司都不以機密處理，並積極提供圖面等大致的架構。所以，在簽訂保密協議之前，至少也要讓 DEF 公司了解此系統大致的情況，以及相對於其它系統的特色等。請您重新考慮，並積極處理。

　　靜候佳音。

【問題點】

　　這篇原文在文法上沒有問題，但是在溝通技巧上卻值得商榷。當人與人溝通時，設身處地為對方著想，是非常重要的。這篇例文似乎就少了這一點；只從自己的立場思考，片面地強調自己的情況。想要說服克拉克先生，卻又一味地強迫他接受我們的商業習慣，恐怕無法獲得對方的認同。

　　我們來看看這篇例文的主要問題點。

　　⑴沒有轉圜的餘地。一味地強調自我的主張，完全不預留讓步的空間。我們可以從最後一句 I look forward to receiving your favorable reply by return. 看到絲毫沒有轉圜的餘地，給人除了 favorable 的回答之外，其餘一概不受理的感覺。這與下述的改善範例 Since time is rather short, may I ask for your response as soon as possible? 就有極大的不同了。另外，像是 by return（立即）這類陳腐的表達方式，實在不討好。

　　⑵不僅是克拉克先生，對多數的歐美人而言，在提供資訊前簽訂保密協議早已是一種共識；如果曾經接獲歐美大型石油公司或建設公司委託正式報價的人，就不難了解這種習慣。在委託

報價之前或同時，他們往往會寄送一份保密協議書，一旦習慣這
種方式之後，估價者就會毫不遲疑地簽字後寄回。順便一提，保
密協議除了 secrecy agreement 的說法之外，也作 confidentiality agreement, confidentiality undertaking, confidential disclosure agreement 等。

例文 1 的改善範例 （○）

① I understand from our New York representative that a problem has come up[1] in regards to the meeting scheduled between yourself and Mr. King of DEF. Our representative reported that you will not meet without a prior secrecy agreement[2] between yourself and DEF. I certainly understand your concern for[3] the security[4] of the New System, and I intend to cooperate with you fully in protecting this technology. As regards a prior secrecy agreement at this stage, however, there are some difficulties which I hope you will let me explain.

② The main problem is the usual procedures now being followed by vendors of competitive systems[5]. It has now become common practice[6] in this market for vendors to make preliminary, non-confidential presentation to potential purchasers[7], in which the outline of the system is described. Purchasers consider this to be a minimum requirement[8] for them to be able to fairly evaluate and compare the various systems available[9].

③ I understand, appreciate[10], and agree with the need[11] for secrecy in the case of this system. I am sure we can protect the secrecy and still maintain practices consistent with[12] the stiff competitive environment[13]. So I sincerely request that you agree to holding the meeting[14] as scheduled, so that we can best use this opportunity to meet a key client[15] in this market.

④ Since time is rather short, may I ask for your response as soon as possible?

1) to come up 發生（問題） 2) prior secrecy agreement 事前簽訂保密協議 3) concern for 關切 4) security （在此表示）防護、保護（＝protection, defence） 5) competitive system 競爭體系 6) common practice 一般習慣 7) potential purchaser 潛在買方 8) a minimum requirement 最低要求 9) available 可獲得的 10) to appreciate 非常了解（＝to understand fully） 11) to agree with the need 同意～是非常重要的 12) consistent with 與～一致、相符合 13) stiff competitive environment 激烈的競爭環境 14) to agree to holding the meeting 同意舉行會議 15) key client 重要客戶

【段落大綱】

① 確認到目前為止的經過，表示了解對方的立場。
② 說明競爭對手一般的推銷方式，並指出買方也認為這種推銷方式是有必要的。

③ 再次表示了解對方立場，同時要求讓步。

④ 促請回答後總結。

【內容】

　　我從本公司紐約代表處得知您與 DEF 公司金先生在安排協商事宜上出了問題。他在報告中說，如果 DEF 公司不事先和您簽訂保密協議的話，您就不打算進行協商。我非常了解您關切新系統的保密問題，而我也願意竭盡全力來保守這項技術的機密。但就現階段而言，事前簽訂保密協議，實在有些困難。請容我在此多做說明。

　　主要的問題在於目前競爭體系賣方所採取的習慣。在這個市場中，賣方對於潛在的買方，事前都會做非機密性的提案簡報，在簡報中說明該系統大致的情形。為了能公正比較、評估各種系統，買方認為更進一步的說明是最起碼的要求。

　　我知道、了解也同意這一系統必須保密，但也相信一定有方法可以兼顧保密，同時符合現今激烈的競爭環境。因此，為了善加利用與此一市場中主要客戶會面的機會，殷切企盼您能按照安排同意舉行會議。

　　由於時間匆促，希望您能儘早回覆。

【注意事項】

　　以上的改善範例，首先表示了解克拉克先生希望事前簽訂保密協議的主張，之後說明一般拓展國內市場的做法，是在不簽訂保密協議的情況之下先做系統概略的說明。然後再次表示能體會克拉克先生的立場，即使不簽訂保密協議，同樣會保守機密，請他務必先與 DEF 公司的金先生會面，以使交易能順利成功。

　　但是，如果克拉克先生認為 DEF 公司（金先生）和 ABC 公司（田先生）站在同一線上，說得難聽一點，就是狼狽為奸的

話，這篇改善範例同樣無法說服克拉克先生。在這個例子之中，DEF 公司應該採取的行動，是依照克拉克先生的主張，簽訂保密協議。

大家不需要認為簽訂保密協議很麻煩，或是覺得對方無法信賴自己而惱怒，應該把簽訂保密協議當作國際商務往來的一個步驟來看待。

2. 不直指對方國名

今日大家都知道我們的工業產品品質優良，獲得全球信賴；但以往本土製品就是低劣產品的代名詞，經常遭受它國恥笑，許多本國企業就曾體驗過這種不平之苦。只要實際體驗過上述情況的人，想必能體會對某國的人說「你們國家製造的產品品質低劣」就等於是傷害這個人。即使情況沒那麼嚴重，但批評某國產品品質惡劣，就等於犯了以偏概全的邏輯錯誤，會造成對方的不愉快。

【背景】

專門銷售機械的 ABC 公司想要將 XXX 國氣閥廠商 XYZ 公司的產品賣給 DEF 建設公司，於是寫了一封信給 XYZ 公司。信的內容是說明如果想拓展國內市場的話，該公司必須改善以往在品質、交貨期和價格方面的做法。

例文 2 促請國外廠商拓展國內市場的信件（原文）

Our client, DEF Construction Company, is now keenly interested in the procurement[1] of plant equipment from overseas countries, including XXX. We at ABC Trading Company, having been in good business relationship with DEF since 1967, would like to assist you to participate as a valve supplier in[2] some of DEF's major projects.

In order to participate in DEF's projects, your company need to be registered[3] in DEF's vendor list. However, we must point out that DEF, based on its experience, has some preconceptions[4] about XXX-made valves. Such preconceptions are: (1) low quality casting[5], (2) unreliable delivery time, and (3) fluctuating[6] prices.

To persuade DEF to include XYZ's name in DEF's vendor list, we would like you to comply with the following three points:

(1) Confirm that XYZ will completely meet the quality requirements of ASME/ANSI B16.34–1988.
(2) Let us know your standard delivery time, from receipt of an order to its f.o.b. delivery[7], that you can guarantee.
(3) Confirm that you will maintain unit prices[8] unchanged throughout the implementation[9] of each individual project.

Could you respond on the above three points by the end of August 199x? Then, we will apply to DEF for[10] approving XYZ as a vendor.

---

1) procurement　購買　　2) to participate in　參加～　　3) to register 登記（參照改善範例註解）　　4) preconception　先入為主的觀念 5) casting　鑄造物　　6) to fluctuate　變動　　7) f.o.b. delivery　（參

照改善範例註解）　8) unit price　單價　9) implementation　執行
10) apply to DEF for　（參照改善範例註解）

【內容】

本公司客戶 DEF 建設公司對於向 XXX 國等海外各國購買建廠機器表示高度興趣。本 ABC 貿易公司自 1967 年以來便與 DEF 公司擁有良好的商務關係，所以希望能在 DEF 公司的幾個主要專案中，幫助貴公司成為 DEF 公司的氣閥供應商。

要參與 DEF 公司的專案計畫，必須先名列於 DEF 公司的採購廠商名單中。但必須說明的是，根據 DEF 公司過去的經驗，他們對於 XXX 國製氣閥有著先入為主的成見，包括：(1)鑄造物品質低劣，(2)交貨期難以信賴，以及(3)價格易變動等。

為了說服 DEF 公司將XYZ 公司列入 DEF 公司採購廠商名單中，希望貴公司能夠遵守下列三點：

(1)請確認完全符合 ASME/ANSI B16.34–1988 的品質要求。

(2)請保證貴公司能夠遵守所告知的從接單到 FOB 交貨的期限。

(3)請確認各專案計畫進行期間，單價不會變動。

請於 199x 年 8 月底前針對以上三點回覆，之後本公司將向 DEF 公司申請，讓該公司認可 XYZ 公司為採購廠商。

【問題點】

這封信中找不到文法上的錯誤。直接點出三項問題點，清楚說明要求的事項，這也是不錯的寫法。如果之前條件不說清楚就開始交易，往往成為日後糾紛的原因。像這封信一樣在交易之前就把基本原則確立下來，可以防止日後發生糾紛。

這封信的目的在於，如果 XYZ 公司對上述三項要求做肯定回覆時，ABC 公司便可將 XYZ 公司積極的態度向 DEF 公司表達，積極向該公司爭取認可 XYZ 公司為採購廠商。如果 XYZ

公司拒絕上述三項要求或是不予回覆的話，ABC 公司可能就會認為不要跟他們往來較好。

　　這種周延的思考雖好，但 ABC 公司有必要多顧慮對方的感受。「根據 DEF 公司過去的經驗，他們對於 XXX 國製氣閥有著先入為主的成見」這句話，如果代換為「本國製氣閥」，恐怕我國的氣閥廠商中也有人會大發雷霆。

　　幸虧最後 XYZ 公司虛心地接受批評。而下述的改善範例中雖然沒有直指國名，但同樣能夠達到要求的效果。如果希望別人接受自己的要求，請勿直指對方國名，因為直指國名，只會傷害到對方的愛國心和自尊心而已。

例文 2 的改善範例（○）

Subject: Participation in Projects of DEF Construction

Dear Mr. YYY:

① ABC Trading Company has worked with DEF Construction since 1967 in the procurement of plant equipment. DEF is now very interested in investigating overseas procurement[1]. We would like, therefore, to invite XYZ to participate as a valve supplier in some of DEF's major projects.

② Such participation will require that XYZ be registered as a vendor with DEF[2]. We must say, however, that as a result of past experience, DEF is cautious about[3] overseas valve suppliers on the following three points: 1) low quality of casting, 2) unreliable delivery, and 3) fluctuating price.

③ If you are interested in participating in DEF projects, may we ask you to do the following?

④ 1) Confirm that XYZ will meet in all respects the quality requirements of ASME/ANSI[4] B16.34–1988.
  2) Let us know the delivery time on f.o.b.[5] basis from receipt of an order that XYZ can guarantee.
  3) Confirm that XYZ will maintain agreed-upon unit prices[6] unchanged throughout any individual project.

⑤ We will then be able to provide reliable reassurance[7] to DEF on these points as we apply for approval for XYZ[8] as a registered vendor.

⑥ May we ask you to respond on these points by the end of August 199x? We would like to make the application for approval[9] following that[10].

---

1) overseas procurement　向海外採購　2) to require that XYZ be registered as a vendor with DEF　XYZ公司必須登記為 DEF 公司的採購廠商　3) be cautious about　對～很謹慎　4) ASME/ANSI　ASME 是美國機械工程師學會（The American Society of Mechanical Engineers）的簡寫，ANSI 則是美國國家標準協會（American National Standards Institute）的縮寫，而 ASME/ANSI 則表示由 ASME 自行擬定的規格經 ANSI 認可者　5) f.o.b.（=free on board）　船上交貨

6) agreed-upon unit prices 協議單價，也作 agreed unit prices
7) reassurance 再次保證，安心 8) to apply for approval for XYZ 替 XYZ 公司申請認可 9) to make the application for approval 進行認可的申請手續 10) following that 之後

【段落大綱】

① 在前言部分，說明公司與客戶之間的關係、客戶的需求和本信函的目的。
② 列舉三個為了達成目的的必要條件。
③ 導入滿足三個必要條件的問題。
④ 將三大要件簡潔而具體地加以陳述。
⑤ 對於問題的回答，預期將可獲得結果。
⑥ 提示回覆期限，說明己方的預定計畫。

【內容】

本 ABC 貿易公司自 1967 年以來就協助 DEF 建設採購工廠設備。目前 DEF 公司對於研究海外採購表示高度的興趣，因此，本公司有意邀請貴公司成為 DEF 公司幾個主要專案計畫的氣閥供應商。

要參與專案計畫，必須先登記成為 DEF 公司的採購廠商。但是就 DEF 公司以往的經驗而言，關於下列三點，他們對海外的氣閥廠商不可謂不審慎。分別是：(1)鑄造物品質低劣，(2)交貨期無法信賴，以及(3)價格的變動。如果貴公司有興趣參與 DEF 公司的專案計畫，請做到以下幾點：

(1)請確認 XYZ 公司符合 ASME/ANSI B16.34–1988 規格的品質要求。
(2)請保證 XYZ 公司能夠遵守所告知的從接單到 FOB 交貨

的期限。

(3)請確認各專案計畫進行中，協議的單價不會變動。

在申請登記 XYZ 公司為採購廠商時，本公司將可針對這幾項提供 DEF 公司可資信賴的再次保證。

以上這三點，請於 199x 年 8 月底前回覆。之後將由本公司代為申請認可。

【注意事項】

(1)改善範例中不曾提及諸如 XXX 國之類的國名，只以海外的氣閥廠商一筆帶過，但卻能與原文一樣傳遞強烈的要求訊息。

(2)原文中先敘述客戶 DEF 公司，之後才介紹自己的公司；在改善範例中則是先介紹自己的公司，接下來才是敘述 DEF 公司。這種先從自己開始介紹的方式應該比較妥當。

(3)相當於「貴公司」的說法，在原文中用了 you 和 your company; 但在改善範例中則使用公司名稱 XYZ。諸如此類在向對方列舉要求事項時，以第三人稱的公司名取代第二人稱的 you 會比較客觀，緩和不禮貌的感覺。

(4)請注意改善範例的客氣與禮貌。在第一段 We would like, therefore, to invite XYZ to participate... 中使用了 to invite 這個非常有禮貌的動詞，而原文則是使用 would like to assist you to participate...這種令人覺得要邀功的用法。第三段...may we ask you to do the following? 是在請求對方許可，而原文則是...we would like you to comply with the following three points...的要求方式。在改善範例最後一段中，以 May we ask to respond...的方式要求許可。有關請求的表達方式，將在第Ⅳ章第 6 節的「請求」中做一整理。

(5)原文和改善範例都是以命令式來表達要求事項。關於命令式的用法，請參照下述的 (參考) 。

（參考）命令式的用法

在日常會話或書信中使用命令式時，不用 please 就顯得很不禮貌，但是在手冊中則不使用 please。

餐廳中侍者問道 Coffee or tea? 時，最簡短的回答就是 Coffee, please. 只說 Coffee. 也能傳達意思，但是缺少 please 就顯得很沒禮貌。

如果想表達「請～」時，也千萬別以為加上 please 就是客氣的說法。譬如書信中如果寫道 Please send it to me by the end of this month. 與其說這是請求，倒不如說是要求。英文中要表達客氣的請求是 Could you ～或 Would you ～或 May I ～等疑問句型，要不就是 I would like ～或 I would appreciate ～等假設用法（參照第 IV 章第 6 節）。「參照～頁」是 Refer to Page ～或 See Page ～，此時不需要加 please。尤其在括弧內或手冊中，絕對不會有 Please refer to Page ～的用法。但是因為我們覺得不寫 please 是很不禮貌的，所以有時會看到極不自然的英文手冊中頻頻使用 please。

在例文 2 的改善範例中，三個要求項目也以命令式表達，這種方式可行嗎? 美國國防部的規格 Military Standard(MIL-STD-490A) 中也有如下的規定:

For specific test procedures, the imperative form may be used provided the entire method is preceded by "the following tests shall be performed," or related wording. Thus, "Turn the indicator to zero and apply 230 volts alternating current."

（在敘述某項特定的測試順序時，可以將「即將執行以下測試」等相關敘述作為先行詞，之後再加上「將指示器歸零，加上交流電 230 伏特」等的命令式。）

　　例文 2 改善範例中的三個項目雖然不是敘述測試順序的文章，但是以此類推，我們先用...may we ask you to do the following? 這種請求許可的表達方式作為先行詞，所以之後可以使用命令式。

3. 捨棄細節，只取大事

　　在交涉重要問題時，可能會出現其它的小問題。交涉者可能會暫時離題，討論這些細節，但是千萬不能因此使得原本最重要的問題處於不利的狀況。為了讓原本重要的交涉問題能處於有利的局面，有時也需要避開細節或讓步。

　　在《孫子兵法》中，有下列名言：

「塗有所不由，軍有所不擊，城有所不攻，地有所不爭，君命有所不受。」

　　一旦開戰，道路也有不通的情況。遇見敵人時，也未必非攻擊不可。當敵人堅守城池時，也未必非攻擊不可。戰場上應當爭取的重要地形或可以佔領的據點，也未必非得佔領不可。即使連君主之命，因時間和場合，也有不能服從的時候。除了最後一句的「君命有所不受」之外，其它都與本例想表達的事情有關。另外，孫子的這段教誨還有下述的英文譯本 (Clavell 1983):

There are roads that must not be followed, towns that must not be besieged. There are armies that must not be attacked, positions that must not be contested, commands of the sovereign that must not be obeyed.

【背景】

　　ABC 公司是進口德國 XYZ 公司的產品到日本銷售的企業，

而 XYZ 公司給 ABC 公司的售價是以日圓計價。

這一年，原本 1 馬克相當於 70 日圓的匯率大幅上揚，變成 1 馬克相當於 90 日圓。為了順應這 30% 日圓貶值、馬克升值的匯率變動， XYZ 公司擬將給 ABC 公司的日圓報價也提高30%，發信向 ABC 公司表明此意（4 月 15 日）。 ABC 公司的鈴木先生反對大幅度調漲，於是寫了份傳真給對方（4 月 26日）。

但是，根據之後拜訪 XYZ 公司回來的 ABC 公司泰勒先生的說法， XYZ 公司的拜爾先生說他並沒有接到那封傳真。甚至拜爾先生還說拍了封電報來要求 ABC 公司同意他在4 月 15 日所提出的調漲要求，但是 ABC 公司的鈴木先生也沒有收到那份電報。

在這種情況下，鈴木先生又寫了一份如下的傳真給拜爾先生。

例文 3　反對漲價的傳真（原文）

TOP URGENT!![1]

Price hike[2] of our XYZ products

Reference is made to Mr. Taylor's meeting with you on 09.05.

Surprised that[3] our telefax[4] A1200 dated 26.04 regarding our proposed CIF prices[5] was not addressed to you, which is enclosed with this telefax[6]. Besides, learnt[7] that you sent a telex on the captioned matter[8] but it is not available for us[9] up to now. Please urgently telefax a copy of the telex to our attention.

Although time has elapsed in vain[10] due to the above un-expected disconnections, we hope our strong and reliable relationship with you doesn't become in danger[11] at all.

We would like you to understand our difficulties of price hike due to yen devaluation[12] in Japan and to accept our proposed prices to be valid from presently pending orders.

Thanking in anticipation[13] and best regards.

N. Suzuki

Enclosure: our telefax A1200 dated 26.04

【問題點】

　　XYZ 公司與 ABC 公司的關係密切，所以鈴木先生對拜爾先生的態度就顯得有話直說。 3)的 surprised 表達出來的是，我們明明傳真過去了，怎麼可能沒有收到的不信任感。 10)的 time has elapsed in vain（白白虛度時光）也讓人有批評對方的感覺。 11)的 we hope our strong and reliable relationship with you doesn't become in danger（希望跟您的深厚信賴關係不會岌岌可危）甚至讓人有要脅的感覺。即使鈴木先生真的有意要脅拜爾先生，達到短期內不漲價的目的，這樣的做法也不會帶來太多的利益。

　　這裡就是犯了沒有好好考慮寫這份傳真目的何在的毛病。下面我們將再加以說明。

1) TOP URGENT!! (十萬火急) 的心情雖然不難理解，但是就傳真的內容而言， URGENT 實在言過其詞。

2) price hike (漲價) 的 hike 用法接近俚語，用 price increase 或 price raise 比較好。

3) surprised that (=I am surprised to know that).

4) facsimile 的語源是拉丁文的 facere (=make, do) 和 simile (=similar, like)，現在英文中已經使用 fax 這個字，另外 telefax 也常用。

5) CIF price (cost, insurance and freight price): 包含運費、保險費的價格。

6) which is enclosed with this telefax enclose 是裝入信封的意思，不能用於「附於傳真之後」的意思。所以最後一行的 enclosure 也不恰當，請參照改善範例。

7) learnt 應該是 we learnt 或 we have learnt，只是省略了主詞。同樣地 3)的 surprised 也是 we are surprised 省略了主詞。這種省略主詞的情況，是用來減少電報中的字數以節省經費，傳真等低成本通訊文章則不必省略。

8) captioned matter: 主題。

9) it is not available for us 的意思等於「電報或許送達了，只是我們沒有拿到」，如果要肯定地說「沒有收到電報」，應該是 it has not reached us。

10) time has elapsed in vain: 白白浪費時間。

11) we hope our strong and reliable relationship with you doesn't become in danger: 希望跟您的深厚信賴關係不會岌岌可危。其中 become in danger 的正確用法應該是 fall into danger。

12) yen devaluation: devaluation 是指固定匯率時政府讓貨幣貶值的意思；如果是浮動匯率的話， devaluation 表示匯率行情下跌 (LDBE)，所以原文的用法是正確的。其它還有 depreciation of the yen, 也是日圓貶值的意思。日圓升值時可用 appreciation

of the yen。簡單的講法，日圓貶值可用 yen's decline, weak yen, low yen rate；日圓升值則為 yen's rise, strong yen, high yen rate。

13) Thanking in anticipation 這種分詞結構的結語已經太古老，不建議您使用這種方式，請參照改善範例。

傳真或電報沒有送達，對 ABC 公司或 XYZ 公司而言都不是什麼大問題；但是價格調漲的問題卻非同小可。鈴木先生應該做的是盡量壓低 XYZ 公司的漲價幅度，但卻挑起沒有收到信件的問題，暗地批評對方，透露不信任的感覺。在開始交涉之前就挑釁，很可能讓交涉變得窒礙難行。

下述改善範例先提到漲價問題，之後才談到沒有收到信件的問題，但並未責難對方。

例文 3 的改善範例（○）

---

1. PROPOSED PRICE INCREASE

① As we discussed in our fax of April 26, 199x, in response to your message of April 15, we understand the circumstances of currency fluctuations[1] which have necessitated an increase in your CIF Yokohama prices[2]. On the other hand, we ask you to understand that, in the severe competitive environment[3] that exists here, we face the risk of considerable loss of share if we attempt to market at the price you have suggested.

② Therefore, we have suggested an alternative[4] set of price increases, which we earnestly request you to consider, in

---

light of[5] the necessity to protect the customer base[6] that has been built upon the work we have done together up to now.　Our customers appreciate[7] the quality of our products, and value[8] the XYZ name.　They also understand the need to take currency fluctuations into account in pricing[9].　We cannot, however, expect to maintain their patronage[10] if the price difference between us and domestic suppliers becomes too great.

③ Naturally, none of our customers like to pay[11] more when our competitors are offering lower prices.　We believe, though, that we can keep customers, even with price increases, if they are on the scale of[12] those we have suggested.　In order to protect and extend the market that has been built upon our mutual efforts, we consider it vital to receive your concurrence[13] with these prices, to be applied to all pending orders[14] and those followings[15].

2. COMMUNICATIONS

④ It is apparent that our facsimiles to you and your telexes to us are somewhat going astray[16].　I would like to confirm these numbers.

(1) Our fax number is xxxxx and our telex number is xxxxx.

(2) Your fax number is xxxxx, and your telex number is xxxxx.

If either of these is incorrect, could you let me know?

Could you reply on this point even if the numbers are correct, so that I can be sure that this message has been successfully transmitted to you?

Best regards[17],

Namio Suzuki

Attachment: copy of our fax message of April 26, 199x (1 page)

---

1) currency fluctuation　匯率變動　　2) CIF Yokohama price　到橫濱為止，包含運費和保險費在內的價格　　3) severe competitive environment　激烈的競爭環境　　4) alternative　替代的　　5) in (the) light of　考慮～　　6) customer base　顧客基礎　　7) to appreciate　肯定～的價值，此時 appreciate 是 to recognize and enjoy the quality 的意思　8) to value　尊重～　　9) pricing　定價　　10) patronage　愛顧，關照　11) none of our customers like to pay　沒有一位顧客願意支付。 none (=not any) 後面接續的 customers 是複數，所以當作複數形處理，但也可作為單數使用　　12) on the scale of　以～的規模　　13) concurrence　同意 (=agreement of opinion)　　14) pending orders　懸而未決（價格未定）的訂單　　15) those following　後續訂單 (=orders which follow)　　16) to go astray　變得行蹤不明　　17) Best regards　用於熟識的對方，此外也常用 Sincerely, Best wishes 和 Cordially 等。

【段落大綱】

① 將漲價問題作為兩家公司必須共同面臨的案件，提出問題點。
② 重複上次傳真中敍述的問題點。
③ 針對上一段中提到的問題提出解決方案，強調解決此一問題是為了雙方著想。
④ 在最後一段談到通訊上的差錯，只當作是撥號的錯誤，並未提及責任的追究。

【內容】

　　1.漲價

　　針對貴公司 4 月 15 日的訊息，本公司在 4 月 26 日以傳真回覆，對於貴公司因為匯率的變動，必須調高橫濱 CIF 價格的情況我們能夠理解。但是另一方面，在日本激烈的競爭環境之下，如果以您所提出的價格來營業的話，恐怕將會失去極大的佔有率，也請您務必諒解。

　　所以，我們提出漲價的替代方案。有鑑於必須鞏固以往共事而打下的顧客基礎，由衷企盼貴公司能夠考慮此一替代方案。顧客們已了解公司產品的品質，也非常重視 XYZ 的品牌。他們也了解設定價格時必須考慮匯率的波動。但是如果我們的價格與國內廠商差距過大，將無法繼續維持客戶對我們的愛護。

　　當然，不會有客戶明明其它業者提出的價格較低而偏偏購買高價的。即便如此，如果漲幅是本公司所建議的，我們有信心保住客戶。為了保有我們共同努力打開的市場，進而擴大佔有率，我們認為貴公司同意本公司的提議價格是非常重要的。此一價格，將適用於現階段價格未定的訂單和往後的訂單。

　　2.通訊

本公司發給貴公司的傳真以及貴公司打給本公司的電報，現在都不知去向。在此確認傳真及電報的號碼。

(1)本公司的傳真號碼為 xxxxx，電報號碼為 xxxxx。

(2)貴公司的傳真號碼為 xxxxx，電報號碼為 xxxxx。

如果其中有任何一個號碼錯誤，能否惠與聯絡？即使號碼正確，也希望能確認貴公司的確收到此一傳真，煩請回覆。

（參考）說服對方與達成協議的談判技巧

歐美的書信或辯論技巧，繼承古希臘、羅馬時代修辭學的傳統。古羅馬時代確立的典型辯論模式如下：

1. 序論 (opening)

2. 陳述 (narrative) 或背景說明

3. 主張 (proposition)

4. 支持主張的理論 (argument)

5. 反論 (refutation)

6. 結論 (closing)

當然，並非所有的辯論都必須按照此一模式推展，可以適度地省略、轉換或擴充。這六個結構要素的核心是 3.主張和 4.支持理論，其它的要素是用來襯托這兩項的。

古羅馬的辯論模式至今仍歷久彌新，例如下列的模式，即為論說式文件的基本形態。

1. 序論 (opening)

1)提出希望引起讀者注意的主題或問題。

2)從筆者的經驗和立場，說明為何能夠闡述此一主題。

3)闡述讀者和筆者共通的信念、立場和經驗，形成與讀者的連動。

2. 背景 (background)

1)陳述問題的本質（經過與原因）。

2)說明問題對讀者的意義。

3. 論說 (argument)

1)闡述（三段論法的）大前提。

2)闡述（三段論法的）小前提。

3)闡述（三段論法的）結論。

4)闡述筆者提案的優越性，主張其它方案無法解決問題。

4. 結論 (closing)

1)說明採用筆者提案的優點以及不採用時的缺點。

2)重複理論、問題與結論。

上述的基本模式對我們而言，感覺上太愛說道理、帶有鬥爭性。我們崇尚「靈犀相通」、「以和為貴」，所以不喜歡也不擅於爭論。

但是，像這篇反對漲價的傳真以及第 1 節例文 1 的傳真都火藥味十足，反而是改善範例比較委婉。正如每一篇的說明中提到，老是主張自己的立場、批評對方的態度、反駁對方的意見，往往會造成反效果。歐美人的溝通，不僅要靠理論推衍來說服他人，還得重視對方的心意，經由刻意的妥協，好讓雙方達成協議。

歐美傳統的說服方式對於信念和價值觀迴異的對手而言，效果適得其反。人們為了求得心中價值觀的安定，對於自己觀念以外的事物都會感到威脅，進而拒絕考慮替代方案。像這種情況，為了使對方讓步，必須降低或緩和替代方案所帶來的脅迫感。根據美國心理學家卡爾・羅傑斯 (Carl Rogers 1902～ 87) 的理論，願意考慮替代方案才是真正溝通的開始，也是解決問題的開端。

類似這種要求同意的文書模式可歸納如下：

1. 闡述問題，表示了解對方的主張。

2. 闡述對方的主張也是正確的。

3. 闡述自己的主張，說明其正確性和適當性。

4. 說明如何採用自己的主張，將對對方有利（雙方的主張如能互補則更為有利）。

這是從 Young 氏的著作 (1970) 中引用下來的。請特別留意例文 3「反對漲價的傳真」和例文 1「促請對方積極拓展國內市場的傳真」的改善範例中，如何應用達成共識的文書模式。

後續連絡與應對

1. 脫離以公司為生活重心的模式

　　一般而言，與歐美人相比，我國的上班族或工程人員顯得工作過量，而且也比較不擅於談論或是書寫除了工作以外的事情。所以即使不在工作崗位上，常常還是會以工作為話題。大多數的歐美人不喜歡在工作以外的時間談論工作，這並不是因為歐美人不愛工作，而是他們有不同的文化和價值觀。

　　下面的例文是一位由公司安排進修的工程人員所寫的感謝函，希望大家也能在寫類似的書信時，從中傳遞出心胸的開闊與人性光輝的一面。

【背景】

　　年輕的工程人員在英國的 ABC 公司接受進修訓練後，寫了一封感謝函給該公司的經理。寫感謝函絕對是件有禮貌的事情。

　　這位工程人員注意到段落要分明，也值得讚許。換句話說，除了稱謂的 Dear Dave 和結尾的 Yours sincerely 之外，還分成四個段落。這種分法很合宜，從視覺上來說相當容易閱讀，但是內容上稍有問題。

例文 4 企業進修生的感謝信（原文）

Dear Dave,

I write to thank you for your kind help you extended during my stay as ABC trainee in your company.

It was indeed a rewarding[1]) and enjoyable[2]) training. I am confident[3]) that the knowledge I learnt through my training will be very useful for the future business activity. I hope that our business relationship established through this training project will also expand to our mutual benefit[4]).

Taking this opportunity[5]), I wish to again convey my personal thanks to you and your staff.

I look forward to our next meeting and I hope it is soon.

Yours sincerely,

1) rewarding　有益的，有利的　　2) enjoyable　愉快的　　3) confident　有信心的（比 sure 文言）　　4) mutual benefit　互利（互惠）
5) taking this opportunity　（請參照改善範例的註解）

【內容】
　　當我以 ABC 公司研究員的身分停留在貴公司進修時，承蒙您親切的協助，特此致函表達謝意。
　　這實在是一次非常有助益而且愉快的進修，我從進修中學習到的知識，相信一定能對未來的工作有所助益。透過這個進修計畫所建立的商務關係，希望能增進雙方的利益。
　　藉此機會，再次表達我個人對您以及您部屬同仁的感謝。
　　希望不久的將來還有機會再見面。

【問題點】
　　第二段...the knowledge I learnt through my training will be

very useful for the future business activity. 或許會給人認真有為好青年的感覺，但是 I hope that our business relationship established through this training project will also expand to our mutual benefit. 未免過於冠冕堂皇。光憑一介工程師前去接受訓練，商務關係就真的能夠建立嗎？而 4)的 mutual benefit（對雙方公司的利益），實在不是年輕受訓人員該有的口氣。

多數的歐美人不喜歡在離開工作崗位之後還談論工作的事情。比如說，在私人的派對中老是三句話不離工作，一定遭人嫌棄。而在生活中總是以企業戰士的精神來面對，也無法融入國際社會。為了能走進世界，見容於國際社會，就必須培養引導話題的能力，增進涵養、判斷力以及語言能力。

總而言之，這篇謝函的問題在於過度偏向公司的立場，應該站在更寬廣的視野和價值觀上，讓精神層面更豐富地來寫謝函。

改善範例也像原文一樣給人企業戰士的印象，但是卻不至於給人一面倒向公司的感覺。

例文 4 的改善範例（○）

Dear Dave,

① Thank you for your kind help extended to me during my stay in London as an ABC trainee. My experience with you and with your colleagues was both enjoyable and rewarding.

② I am confident that the new skills I was able to gain[1] while with you will prove useful[2] to my company, and I hope that I will also be able to contribute to the relationship between your company and mine.

③ May I also take this opportunity to[3] convey my deep appreciation[4] to your staff for the invaluable[5] assistance which they extended to me during my stay.

④ I look forward to the chance to meet with you all[6] again.

Yours sincerely,

1) to gain　獲得（努力的結果），比 obtain 稍微正式　　2) to prove useful　證明（結果）～是有用的　　3) to take this opportunity to do (or of doing)　利用這個機會做～　　4) appreciation　感謝（appreciation 有「真正了解」的意思）　　5) invaluable　無比寶貴的 (=extremely useful)　　6) to meet with you all　（請參照注意事項(3)）

【段落大綱】

① 先表示謝意，敘述進修有趣而且收穫豐碩。
② 表示希望未來能發揮在進修中學習到的技術。
③ 表達對對方部屬的謝意。
④ 以期待能再見面作結。

【內容】

　　當我停留倫敦，在 ABC 公司進修時，非常謝謝您的幫忙。跟您及您的同事共處的這段經驗，實在非常快樂而且深具價值。

　　我相信，在貴公司停留的這段期間中學習到的新技術，一定對本公司有所助益，而且也能對貴我雙方公司的關係做出貢獻。

　　藉此我要表達內心誠摯的謝忱，謝謝在我停留的這段期間，

您的部屬所給與無比寶貴的相助。

期待下次再與各位見面的機會。

【注意事項】

⑴第二段中間,「本公司」的英文是 my company 而非 our company; 第二段的結尾則用 mine。因為 our company 有「貴公司和本公司」的意思,所以在說「本公司」的時候,應該用 my company。

⑵原文開頭先對達夫言謝,之後再對達夫和他的部屬表示感謝。而在改善範例之中先對達夫表示感激之後,第二次就只感謝他的部屬。改善範例比較簡潔乾脆。

⑶原文結尾為 I look forward to our next meeting...感覺是要和對方約定下次的見面,但其實並不是要作任何約定,所以改善範例中改成 I look forward to the chance to meet with you all...比較好。

2. 保持一貫的語氣

「先從好消息開始寫,壞消息要寫在好消息之後」,這是商業書信的原則,因為從好消息開始寫的話,讀信的人比較容易接受信件的內容。但是,如果好消息和壞消息的落差太大,反而會讓信函變得很唐突。

【背景】

ABC 公司海外採購課的楊經理接待英國 XYZ 公司史密斯先生的來訪。 XYZ 公司具有地下埋設管線的檢查技術。由於現場的技術問題不在楊經理的負責範圍之內,於是他把 XYZ 公司留下的資料轉給維修部經理。此外,在會談中,又因為印度埋設的地下管線可能發生外漏的問題,於是也向史密斯先生提及想將

XYZ 公司介紹給 ABC 公司孟買事務所的李先生。

　　不久之後，楊經理接到史密斯先生例文 5 的信件，得知 XYZ 印度分公司要與孟買事務所連絡而大為緊張，因為之後他聽維修部經理報告說， XYZ公司的技術並不適用於印度所發生的外漏問題。

　　例文 5 是 XYZ 公司史密斯先生的來信。業務人員在拜訪過客戶之後，寫信致謝是理所當然；但這封信不僅僅是致謝，還針對拜訪時談到的話題加以確認，並對客戶所提到的問題（外漏）提出解決方案，相當具有積極性。

例文 5 訪問後的後續信件（○）

Dear Mr Young

① I am very glad to have had the opportunity of meeting you and your colleagues to present[1] XYZ services and products to you on 22nd March.

② One of the matters raised[2] was the problem of leaks in underground pipelines. In particular, India was mentioned as a country where ABC could be interested in assistance from XYZ in inspecting that type of leak.

③ Our Indian Company, XYZ India, has technicians[3] trained in our pipeline inspection techniques, and we have asked our General Manager there, Mr D Gupta, to contact[4] Mr K Lee at your Bombay office to discuss possibilities of cooperation[5].

④ If you should require XYZ's services or products at any of your bases around the world, please contact us and we would be glad to make a rapid response[6] to your requests.

Yours sincerely

Donald Smith
Director

1) to present　進行簡報提案、介紹　2) to raise　提出（問題等）
3) technician　擁有特殊技術的專業人員　4) to contact　「連絡」
（及物動詞），例如：For further information, please contact our sales office.（詳情請洽本公司營業部。） to contact with 的 contact 是不及物動詞，表示「接觸」　5) cooperation　合作　6) rapid response
迅速的回應

【段落大綱】

① 感謝對方給予簡報提案的機會。
② 重複協商中提到的問題。
③ 本公司能處理該問題，並正準備採取行動。
④ 若有其它地區發生問題，也懇請連絡。

【內容】

　　感謝您讓我有機會在 3 月 22 日造訪您及您的同事，並介紹 XYZ 公司的服務與產品。

　　當時曾提及關於地下埋設管線的外漏問題。其中特別提到印度 ABC 公司可能會對 XYZ 公司協助檢查外漏感興趣。

　　本 XYZ 印度分公司擁有受過管線檢查訓練的技術人員，於

是我請該公司總經理葛普塔先生與貴公司孟買事務所的李先生連絡，洽談合作的可行性。

　　貴公司全球各地的任一據點如果有需要 XYZ 公司的服務或產品時，請惠與連絡。我們將立即回應您的需求。

【注意事項】

　　在拜訪客戶之後寄發謝函，同時確認協商的內容，並說明已經具體指示印度相關公司與孟買方面連絡（第三段）。我們的說法是「指示」或「命令」印度公司，因此就會使用 instruct 或 order。但是請注意，史密斯用的是「請求」(ask)，把葛普塔先生列於與自己對等的地位。史密斯先生是母公司的董事，而葛普塔先生只是子公司的總經理，兩者間地位的落差明顯；更何況，印度還是以前英國的殖民地呢！

　　史密斯先生對於子公司、而且地位比自己還低的葛普塔先生以相等的地位對待，其實是希望有助於葛普塔先生的業務。

　　還有一點，第二段 India was mentioned as a country where ABC could be interested in assistance from XYZ...，如果換作 You mentioned India as a country where ABC could be interested in assistance from XYZ...（您提到印度 ABC 公司可能會對 XYZ 公司的協助感興趣）主動式的話，會有什麼不同呢？考慮前後文的狀況， ABC 公司楊經理的確提到印度管線可能有外漏的疑慮，但卻不是非常肯定。所以在此用被動式表達，不明確指出是誰說的，這也是史密斯先生考慮到減輕楊經理負擔的用心。

　　這封信有幾個英國式的特色，不同於美國式。開頭是 Mr Young（非 Mr. Young）、第三段的 Mr D Gupta （非 Mr. D. Gupta），英式英文中會省略句點。第一段的 22nd March，美國人可能會寫成 March 22。另外，開頭稱謂的 Dear Mr Young 後不打逗點（英式）或冒號（美式），結尾的 Yours sincerely 也不打逗點，這是英國常見的開放式標點符號（open punctuation）。

例文 6　企業拒絕訪問的信件（原文）

Dear Mr. Smith,

Thank you for your letter of 20th April. Immediately after you left our office on 22nd March, I conveyed the brochure received from you to our Maintenance Department.

Mr. S. Mason, Maintenance Manager, well knows about the reputation of XYZ and intends to contact XYZ Japan whenever he needs XYZ services.

Concerning underground pipeline inspection in India, I think we must approach this problem differently from your method. Therefore, please instruct Mr. D. Gupta not to spend time by visiting our Bombay office.

I look forward to having opportunity of working with you.

Yours sincerely,

K. Young, Manager
Procurement Section

【內容】

　　謝謝您 4 月 20 日的來信。當您於 3 月 22 日離開本公司之後，我就把收到的簡介轉交給維修部門了。

　　維修部經理梅森先生深知 XYZ 公司的口碑，所以需要 XYZ

公司的服務時，將會隨時連絡日本的 XYZ 公司。

　　至於印度埋設的地下管線檢查問題，因必須使用和貴公司不同的方法來處理，所以請指示葛普塔先生無須浪費時間前往本公司的孟買事務所。

　　期待今後有共事的機會。

【問題點】

　　這封信的問題點在於第一段和第二段都很順，可是第三段卻斷然地拒絕別人的拜訪，使得兩種不同的語氣混在一起。因此前半段的好消息，以及 I look forward to having opportunity of working with you. 的美好結尾，都只給人做作的感覺。

　　第一段表示在史密斯先生回去後，已經採取行動，讓史密斯先生深感高興。第二段也提到知道 XYZ 公司的名字，寫到這裡一切都很好。但是接下來的內容，有必要時會連絡日本 XYZ 公司，換句話說，你不用再來推銷了，等於開始架設防護網。

　　第三段就顯得相當性急。拒絕葛普塔先生來訪的用語，不僅片面地反駁，而且...please instruct Mr. D. Gupta中用 instruct（指示）這種上對下的動詞，就我們的感覺而言，可能是對史密斯先生的敬語，但也可能被認為是階級意識的表現。請注意史密斯先生用的是沒有階級差別的動詞 ask。而 to spend time 與 to waste time（浪費時間）一樣，等於斬釘截鐵地說「來了也是沒用」。

　　這封信的問題點在於前後的語氣差異太大。換個角度來看，前半段和後半段的主題不同，一封信裡有兩個主題，就會出現問題。一封信只要一個主題就好。

　　下面的改善範例同樣是拒絕葛普塔先生的拜訪，但是拒絕的方式比較自然。

例文 6 的改善範例（○）

---

① Thank you for your gracious[1] letter of April 20. May I assure you that[2] we are aware of[3] the excellent reputation of the XYZ organization; in fact, we intend to contact your Japanese associate[4], XYZ Japan, to obtain these services when it is necessary.

② As for the question of leaks in India, I am informed that an approach different from the XYZ one is planned, so while[5] we appreciate the kindness of your concern, it appears that this would not be an opportune[6] time for a visit from Mr. D. Gupta. Perhaps some time in the future, when an appropriate[7] project presents itself[8], we might be able to call upon him for his services[9].

---

1) gracious=polite　有禮貌的　　2) to assure you that　向您保證～
3) be aware of　知道（=have knowledge of），類似的表達方式有 be conscious of（自己知道～）　　4) associate　合作夥伴（公司）
5) while　表示稍微的對立（=on the other hand）　6) opportune　（時間上）合宜的　　7) appropriate　適當的（=suitable）　8) to present itself　顯現　9) to call upon (or to call on) him for his services　請求他的服務

【段落大綱】

① 感謝對方來信，給予對方高度的評價，表示有需要將隨時連絡。

> ② 但是有禮貌地拒絕，説明現階段的問題須以其它方法解
> 決。

【內容】

　　謝謝您 4 月20 日誠摯的信函。我們深知 XYZ 公司優良的口碑，當有需要時，我們將會連絡貴公司的合作夥伴日本 XYZ 公司，接受他們的服務。

　　關於印度的外漏問題，聽說是計畫以不同於 XYZ 公司的方法來進行，所以我們很感謝您的關心，但目前似乎不是葛普塔先生來訪的好時機。未來如果有合適的案件，屆時希望能夠麻煩葛普塔先生為本公司服務。

3.　協議記錄應當場記錄

　　一般而言，與客戶或承包商協商後，都會以書面再次相互確認同意事項和協商的內容。以書面確認協商結果的方法，有下列三種：

　　⑴在協商現場就作好協商記錄（minutes of meeting 或 memorandum 或 meeting memo），由出席人員簽名。

　　⑵協商後，由一方作好記錄送交對方，請對方在記錄上簽名後寄回。

　　⑶協商後，由一方以信件的方式，寫明協商內容及同意事項，請對方確認。

　　如果協商內容和同意事項非常重要的話，就應該以⑴的方式，在當場就作好協商記錄，彼此確認後簽字。在當場沒有時間作好記錄，雖然也可以採取⑵的方法，但是如果記錄的內容對方有異議，就等於要重新討論。所以，與其選擇⑵的方法，倒不如

加把勁,採用(1)的方法,在當場就做好確認的工作。如果協商的
內容並不重要,就可以採用(3)的方式,以書面作單向確認即可。
雖然對方如果不以書面方式作確認的話,這封信並不具效力,但
總比沒有任何記錄要來得好。

【背景】

楊經理很早以前就向客戶 XYZ 公司提議修繕管線生產設
備,在這次的協商中,終於讓對方口頭承諾要正式地進行。回國
後,楊經理寫了下面的這封信給客戶 XYZ 公司的技師長貝克。

例文 7 協議後的後續信件 (原文)

January 30, 199x

Messrs. XYZ Steel Pipe Company[1]

Attn: Mr. L. M. Becker
      Chief Engineer

Dear Sirs;[2]

Re: Plant Revamping Project

It is our great pleasure to have had a fruitful discussion[3]
for the captioned project with you and other personnel[4]
concerned on 27th this January at your office, and to have
heard that the official inquiry for the project will be issued
quite soon.

Taking this opportunity, we would like to confirm the result
of the said meeting[5] as follows:

1) Preliminary discussion was held on certain key points of "Technical Specification" prepared by you in order to clarify the contents thereof[6] and XYZ's intention.

2) Regarding the automatic welding machines referred to under Para. 4.4 of the Technical Specification, its list with requirements will be submitted to us by February 10, 1991 through your head office.

We would like to inform you hereby[7] that, as the result of preliminary study of the Technical Specification at our head office, we are basically able to follow up[8] your requirements and are ready to start proposal work immediately after receiving your official inquiry. Expected time schedule for materialization of the project is attached hereto[9] for your reference.

We sincerely wish[10] that this epoch-making[11] project will be successfully realized and completed under our mutual collaboration.

Yours faithfully,

_____

J. Young
Proposal Manager

Confirmed by:

_____

for and on behalf of XYZ

Attached: Proposal Schedule

【內容】

　　XYZ 鋼管公司

　　L.M.貝克技師長鈞鑒

主旨: 工廠修繕工程

　　非常高興今年 1 月 27 日於貴公司與您及其它人員針對上述提案有成果豐碩的討論，並得知在不久之後，本案將進入正式的報價程序。

　　利用這個機會，跟您確認上次的協商結果。

1) 為了確認您製作的「技術規格書」內容與貴公司的意願，於日前討論了該規格書的要點。

2) 關於技術規格書 4.4 項中所提的自動焊接機，將於 2 月 10 日以前，透過貴公司總公司提出一覽表與要求事項給本公司。

　　我方總公司在事前討論過技術規格書之後，原則上將可遵照您的要求，特此通知您。如果接獲正式委託，我們已經準備好立即開始提案作業。茲附上本案的實行預定時間表，請參考。

　　由衷企盼此一劃時代的案件能夠順利實現，且在雙方的合作下完成。

企畫經理　　J. 楊　敬上

簽名:

謹代表 XYZ 公司確認上述內容。

附件: 提案預定表

【問題點】

　　看看這封信的左下方結尾處，預留有客戶 XYZ 公司代表的簽名欄。楊經理大概是希望客戶中有人 (貝克技師長等) 能在此

簽名，確認協商內容吧！但是，與其用 for and on behalf of XYZ 這種合約式的方法請對方簽字，還不如在協商當場就做好協商記錄，然後雙方簽字確認。更何況，看看這封信的內容，似乎也沒有重要到必須 XYZ 公司簽字的程度。

還有，這封信給人一股傲慢的感覺，客戶和賣方的立場似乎正好相反。我們來看看細節的部分，找出原因所在。

1) Messrs. XYZ Steel Pipe Company：在公司名稱前加上 Messrs.，現在只用於公司名稱是以個人名字為開頭、個人色彩較強的公司。

2) Dear Sirs;的分號（；）：由於這是一封英式書信，所以應該用逗點，美式書信才是用冒號（：）。

3) a fruitful discussion（成果豐碩的討論）：雖然能夠了解筆者想要說什麼，但 fruitful 給人浮華做作的感覺。

4) personnel 指的是公司全體員工，沒有單指一個人的感覺，所以不適用。這裡用 your staff（部屬）或 your colleagues（同事）比較好。

5) the said meeting 的 said 是法律用語，同樣的 6) thereof、7) hereby、9) hereto 等也都是法律用語，不適用於這封信。

8) we are basically able to follow up 與 maybe we are able to follow up 同樣給人否定的感覺，楊經理應該不是否定的意思，而只是小心謹慎罷了吧！

10) We sincerely wish 的 wish 原本是有期待不可能實現的事情發生的意思，而 hope 則是表示有可能實現的願望。

11) epoch-making「劃時代的」未免過於誇張，這裡實在不適用。

例文 7 的改善範例（〇）

Dear Mr. Becker:[1]

Re: Confirmation of Discussion, January 27, 199x
    Pipe Fabrication Plant Revamping[2] Project

① Following upon[3] our meeting of January 27, we would like to confirm that you share our understanding[4] of discussions at that meeting:

② 1) Agreement was reached[5] on the content of Technical Specifications in order to meet requirements for the Project specified by XYZ.

2) XYZ Head Office will forward[6] to us by February 5, 199x a list of requirements for the automatic welding machines[7] described in Para.4.4 of the Technical Specification.

③ It was a pleasure to be able to meet you and your colleagues during these discussions, and we were gratified to hear[8] that you will be issuing your Official Inquiry[9] to us soon. On our part, our Head Office, having studied the Technical Specifications agreed upon at our meeting, reports itself ready to begin preparation of a Proposal immediately upon receipt of your Official Inquiry. For your reference[10], I also enclose a draft[11] schedule for completion of this project.

④ Our entire company looks forward with enthusiasm[12] to collaborating with[13] you and your staff in the successful realization of this project.

Yours very truly,

John Young
Proposal Manager

Encl.[14]

1) Dear Mr. Becker 比 Dear Sirs 要來得沒有距離，也比較自然　2) revamping　修繕　3) following upon　隨著～之後　4) to share our understanding　分享我們的了解（相同的看法）　5) to reach (an) agreement　達成共識，同意　6) to forward　送（比 to send 生硬）　7) automatic welding machine　自動焊接機　8) we are gratified to hear　很高興聽到　9) official inquiry　正式詢價、交易　10) for your reference　供您參考　11) draft　草稿、草案　12) with enthusiasm 熱烈地　13) to collaborate with～　與～一起工作 (=to work together with)　14) Encl. Enclosure 的縮寫，表示附上文件之意。文件的名稱已經在本文中出現過了，在此省略。

【段落大綱】

① 說明本信的目的在確認協商的結果。
② 將協商的結果整理為二項。
③ 表達期待對方提出正式詢價，並說明信中附上計畫時程表。
④ 表達期待本案的執行並結尾。

【內容】

主旨：鋼管生產工廠修繕工程
　　　199x 年 1 月 27 日協商內容的確認

希望能確認貴我雙方於 1 月 27 日協商中討論的結果，雙方已經達成共識。

1) 為了滿足 XYZ 公司對本案所提出的要求事項，就技術規格書內容達成協議。

2) XYZ 公司總公司於 2 月 5 日之前，將技術規格書中第 4.4 項自動焊接機相關要求事項一覽表寄給我方。

非常高興能夠在協商的過程中見到您及您的同事，並得知正式詢價文件即將發出，令人十分欣喜。我方總公司已經研讀過協商時所同意的技術規格書，等接獲貴公司的正式委託之後，立即開始提案作業。另附上本案時程表以供參考。

本公司全體同仁熱切期待和您及各位合作，成功實現本案。

【注意事項】

信的前半段在確認協商的同意內容。原文 1) 中只有提到「討論」 (discussion)，而改善範例則以「同意」 (agreement) 的字眼來作具體的說明。後半段表明本公司對此案的積極態度，將段落整理得乾淨俐落，並以全公司的期待作結尾。

　　因為內容並非須獲得 XYZ 公司的確認不可，所以沒有像原文一樣留有 XYZ 公司的確認簽字欄。請注意信件標題提綱挈領，非常具體。

通知與回覆

1. 致歉前先考慮

我們經常沒做錯什麼就「對不起」、「對不起」一直掛在口邊，但是大家也知道，歐美人是不會隨便道歉的。不過話又說回來，歐美人也常說 Excuse me 或 Sorry，從這點看來，歐美人和我們對何時該道歉或許有不同的看法，但是勇於道歉的精神則是一致的。無論如何，當我們想要道歉時，最好先想想接受訊息者聽到道歉時會有什麼樣的感受。舉例而言，在交易上為了小過失而致歉時，就得先考慮是否有發展為賠償損害的可能性。

下面是一封董事異動通知書，對延遲告知表示歉意。我們來看看這封信是否合宜。

例文 8 高階主管人事異動的通知（原文）

> This is to formally inform a personnel change in our service to XYZ.
>
> Mr. Paul Lee, Director and General Manager of Technical Division, has replaced Mr. Young as the Director for XYZ and the signature on the Agreement for 1995–1998 enclosed here was executed by Mr. Lee.
>
> I apologize for delay in sending the Agreement and your kind understanding and cooperation with this change will be very much appreciated.

【內容】

特此正式通知本公司負責貴 XYZ 公司業務的主管人事異動。

XYZ 公司的負責主管由董事暨技術部總經理李保羅先生取代楊先生，附在信內 1995–1998 年協議書上的署名為李先生的簽字。

很抱歉延遲寄送協議書。請諒解此次人事異動，非常感謝您的合作。

【問題點】

第三段針對延遲寄送協議書表示歉意，對我們而言，是非常理所當然的禮貌。但是歐美人的觀念不太一樣，當然致歉的方式也有差異。不過對歐美人而言，道歉等於承認自己的不是，所以一般而言，是不會輕易道歉的。

第一句 to formally inform 給人的感覺是以前已經非正式地告知過，這次是正式的通知。

最後一句除了用 apologize 不當之外，your understanding and cooperation will be very much appreciated 用被動式也讓人覺得奇怪，因為這樣會引起 By whom? 的疑問，所以一般應該寫成 We would very much appreciate ～。

例文 8 的改善範例 A（以主管人員替換為主題）（○）

① This is to formally announce that Mr. Paul Lee, Director and General Manager of Technical Division, has been appointed Board Representative[1] in charge of[2] relations with XYZ, succeeding Mr. Young.

② The signature on the Agreement for 1995–1998 is that of Mr. Lee. The execution of this Agreement[3] was delayed pending[4] Mr. Lee's appointment, but I am happy to be able to enclose it.

1) Paul Lee...has been appointed（可以加上 as 或 to be） Board Representative　李保羅被指派為董事代表　　2) in charge of　負責～
3) execution of this Agreement　於協議書上簽名。如果是 execution of the project，則是計畫的執行　　4) pending　（介系詞，文章用語）＝until

【段落大綱】

① 告知主管人員的人事異動，以及新董事的名字。
② 敍述附上新主管簽名的協議書。

【內容】

　　在此正式通知，董事暨技術部總經理李保羅先生接替楊先生，被指派為 XYZ 公司相關業務的董事代表。

　　1995–1998 年協議書上的署名是李先生的簽字。這份協議書的署名一直延遲到李先生被指派為止。特此附上。

【注意事項】

　　相對於原文中對延遲簽名表示歉意，改善範例 A 說明了延遲的理由，並將寄送協議書當作是一個好消息。此外，原文中用李先生 replace 楊先生，請注意，在改善範例中則是以李先生被指派的說法。即使是董事換人，也不能說董事「取代」，這時用「被指派」比較謙虛；而且事實上也並非自己就任，而是被指派

的。另外, replace 有「更替」的意思。

例文 8 的改善範例 B (以寄送協議書為主題) (○)

① I am pleased to be able to send you the Agreement 1995–1998.

② The execution of this Agreement has been delayed pending the assignment of a new Board Representative in charge of relations with XYZ. I am pleased to be able to inform you that, as of April 10[1)], Mr. Paul Lee, Director and General Manager of Technical Division, was appointed to this position, succeeding Mr. Young. You will notice[2)] that this Agreement has been executed by Mr. Lee's signature.

③ Mr. Lee, and all of us, look forward to continuing the warm relationship that we have enjoyed with XYZ.

1) as of April 10　4 月 10 日的　　2) to notice　注意～, 同樣也可以使用 to note

【段落大綱】

① 提到寄送協議書。
② 敘述董事換人, 協議書是由新董事簽字。
③ 對未來表示積極的態度並作結。

【內容】

　　很高興能在此將 1995–1998 年的協議書寄上。

　　這份協議書的簽字，由於受到負責 XYZ 公司相關業務的董事代表異動而延遲， 4 月 10 日，董事暨技術部總經理李保羅先生接替楊先生，受指派擔任此一職務，特此奉告。相信您也已經注意到，此協議書的署名乃李先生的簽字。

　　李先生暨全體員工共同企盼保持與 XYZ 公司的良好關係。

【注意事項】

　　改善範例 B 是以寄送協議書為主題，附帶地提供人事異動的消息。與改善範例 A 一樣，並沒有對延遲寄送協議書而道歉。

> （參考）致歉的表達方式

　　歐美人不像我們那麼容易道歉，但是英文裡也有致歉的表達方式，而且日常生活中也會用到。我們來看看含有道歉意思的表達方式是如何使用的。

regret（及物動詞）：即 to feel sorry about，表示「後悔、可惜～」的意思，用於傳遞壞消息或有禮貌地拒絕時。

We regret to inform you that you owe the bank $100.

（很抱歉，特此通知您欠銀行 100 元。）

I regret that I will be unable to attend.

（很可惜，我無法出席。）

sorry（形容詞）：表示「很可憐、覺得難過」。

I'm sorry to have kept you waiting.

（抱歉讓您久等。）

　　曾經從德國某公司接到下面的一份傳真，有如下的「對不起」表達方式。

We are sorry to inform you that due to the shortage of designers we have at this moment because of the vacation period, we are unable to support you with our price info.

（由於正值休假期間，目前設計師不足，很抱歉無法協助提供價格資訊。） [info 是 information 的縮寫，但這不是電報，無須縮寫。]

另外，下面這篇範例是荷蘭某公司針對對方的行為表示遺憾的 sorry，換句話說，就是在指責對方。

We are sorry to understand that you have the intention to place an order with one of our competitors.

（我們了解貴公司有意向本公司的競爭對手之一下訂單，實在令人遺憾。）

excuse（及物動詞）「原諒」： Excuse me.（對不起）用於跟陌生人說話時、碰到別人、經過他人面前等時候。下面是一封英國公司打來的電報。

EXCUSE DIRECT CONTACT. I HAVE BEEN GIVEN YOUR NAME BY MY COLLEAGUE,...

（請原諒我直接拍電報。我從同事處得知您的大名……）
pardon 的用法與 excuse 相同。

forgive（及物動詞）「原諒」： 與 excuse 和 pardon 相同，但 forgive 的情況較嚴重。

apologize（不及物動詞）「抱歉」： 這個字的語源是希臘文的 apologia，是由 apo（離開）和 logos（語言）組成，表示為了脫罪的語言；換句話說，就是為了保護自己所作的辯解。柏拉圖記錄蘇格拉底接受死刑宣判時的辯論，那篇記錄叫做 *Apology*《辨明》相當著名。當然，apology 現在也還有「辨明」(a defence or explanation of a belief, idea, etc., *LDCE*) 的意思。同樣的，

excuse 也有「狡辯」的意思，名詞形的excuse 第一個意義就是「藉口」。

apologize 的用法有：

I must apologize for not replying sooner to your letter.

（很抱歉無法更早回覆您的信件。）

用法與「久未問候，尚請見諒」相同。

下面是一封實際從美國寄來的信件部分內容。

I have to make an apology to you. Item 1.1 of my last fax to you is in error.

（我必須向您道歉。我上次傳真中 1.1 項有誤。）

這裡的 error 並非重大錯誤，所以 apology 也可以用於輕微的道歉。但是下面的 apologize 在商務上則有重大意義。底下這封是另一家美國公司寄來的信件。

We wish to apologize for the lateness in shipment. This delay was caused by the late delivery of castings to our manufacturing facility.

（貨物延遲裝船，特此致歉。延誤的原因在於鑄品交貨給本廠的時間延遲所致。）

這麼坦誠的道歉方式，在歐美的商業書信中並不多見。可能是這家公司基於與我國企業交易的經驗，猜想對我們最好老老實實地道歉，會給對方好感；就算認錯，對方也不會索賠，對未來的交易只有正面的影響。

像這種情況，歐美企業一般的反應如下：

We regret the lateness in shipment. However, this delay was beyond our control because it was caused by the late delivery of castings to our manufacturing facility.

原料延遲交貨才是原因，這是本公司無法控制的 (beyond our control)，換句話說，就是主張乃不可抗拒 (force majeure)

的因素所致。歐美的合約書中大都會註明，因不可抗拒之因素而
延誤時，不得請求支付違約金。

如同上述內容，歐美人的商業書信中也有致歉的表達方式，
有時則是禮貌性地表示遺憾或歉意。在指責對方時，也有以表達
遺憾來達到目的的方式。

在想以我們的方式道歉時，應該注意先從對方的立場，仔細
思考是否非道歉不可，不需要讓自己一直處於不利的立場。

2. 有人情味的短箋

與海外公司合作的計畫，會依據相當於計畫憲法的合約（con-
tract 或 agreement）、合約的附件（annex 或 appendix 或 ex-
hibit）、或是履行合約過程中訂好的商議記錄 (minutes of meeting) 來執行。因此，當事者之間的商業書信往往是以這些文件
為引用的根據，這也使得商業書信流於古板，文字顯得沒有人情
味。

下面的例子是在告訴對方自己公司負責人的名字，同時詢問
對方（客戶）負責人的姓名。

例文 9　負責人姓名的通知（原文）

Re: Notification of the Seller's Representative at XYZ Construction Site

It is our greatest pleasure to inform you that Mr. John Young,
who will be the chief engineer of the Seller at XYZ construction site, is assigned as the Seller's official representative at

the construction site, in accordance with Par.10–2 of Appendix 6 to the Contract.

In the meantime, according to the same paragraph mentioned above, we would like to be advised in writing of the name of the Buyer's representative at the same construction site.

【內容】

主旨：　XYZ 工地賣方代表人的通知

特此通知楊約翰先生依據合約附件 6 之10–2 項，成為工地賣方正式代表，擔任 XYZ 工地賣方的技師長。

此外，依據上述同一條款，請以書面告知該工地買方代表人大名。

【問題點】

這篇例文有三處語氣上缺乏人情味。第一是合約附件條款的引用。如果是這種類型的單純事項，其實也用不著引用合約的條款。第二處是稱對方為 the Buyer，稱自己公司為 the Seller。合約中為避免不斷重複買賣雙方的名字或簡稱，常以 Buyer （或 Purchaser, Owner）與 Seller （或 Contractor, Vendor, Supplier）來稱呼，但信件中不須使用這種稱謂。第三處是 is assigned as...（第二段）和最後一段的 we would like to be advised of...，被動式使用頻繁。尤其是最後一段，用主動式會比較自然。

例文 9 的改善範例（○）

Re: Appointment of Official Representatives, XYZ Construction Site[1]

① As required by Paragraph 10–2 of Appendix 6 to the Contract, we have appointed Mr. John Young, Chief Engineer[2], as[3] the official representative of ABC Corporation at the XYZ construction site.

② In accordance with the same paragraph of the Contract, may we also have your written notice[4] of the appointment of your official representative to this construction site?

1) construction site 工程現場、工地　2) chief engineer 技師長、工程組長　3) to appoint ～as... 指派～為……(=to assign)
4) written notice 書面通知

【段落大綱】

① 引用合約附件條款，告知負責人姓名。
② 詢問對方負責人姓名。

【內容】

　　主旨: XYZ工地正式代表的指派

　　基於合約附件 6 之 10–2 項要求，指派本公司技師長楊約翰先生擔任 XYZ 工地 ABC 公司的正式代表。

　　依據合約同一條款，希望獲得貴公司工地正式代表的任命通知書。

【注意事項】

　　改善範例中沒有使用 the Buyer, the Seller，與原文不同，但是仍有引用合約條款。最後的 may we also have...是有禮貌地請求。

3. 傳達好消息

　　當買方要變更技術規格時，無論對方是在國內或海外，千萬不可忽略書面的變更手續。一旦輕忽了這個手續，可能會造成規格變更的連絡失誤，或是延誤交貨、漲價等情況。所以，確實處理規格變更，是順利進行交易的重點之一。

　　通知規格變更時應注意，千萬不能因過度顧慮規格變更既耗時又會影響計畫的心理，而使規格變更成為一個壞消息，如此會使對方的反應變得更負面。所以在傳達帶有負面因素的設計變更時，要用點腦筋，在措詞上用點技巧。當然，大前提還是要正確的傳遞，不能產生誤會。

　　以下介紹的例子，是某建築設計事務所的傳真。原本受客戶委託在海外興建的建築物，由於規格變更，使得原本當地的建築業者仍在估算建築工程款項時，就必須重新針對新規格進行估價。該建築設計事務所就是以這份傳真來通知當地建築業者的。

例文 10 建築物規格變更的通知 (原文)

Please be informed that the following revision has been issued by the customer. We would like to ask you submit your quotation accordingly.

- Unit air-conditioning system for the building has not been approved. Please cancel the description Para. 2.2 of our specification. Unit air conditioners shall be excluded from your supply scope.

- Instead of the unit air conditioning, central air conditioning shall be employed. The conditions shall be per the attached sheet. The whole central air-conditioning system shall be included in your supply scope.

We regret to inform of this revision but hope to get your firm quotation on schedule.

Your kind cooperation will be highly appreciated.

【內容】

　　茲通知下述變更乃是由客戶所提出。請依據此通知，提出估價單。

　　建築物的單位式空調系統沒有獲得認可。請取消本公司規格書中之 2.2 項。貴公司供貨的範圍內，將刪除單位式空調機器。

　　將採用中央空調以取代單位式空調，條件如附件所示。整個

中央空調系統將列入貴公司供貨範圍之內。

很抱歉通知您這項變更，希望按照原定行程接到貴公司確定的報價單。

感謝您的合作。

【問題點】

這份傳真的口氣是將規格變更當作壞事來傳遞。不要把規格變更當成壞消息，要以更正面的口吻來書寫，並應明確地指示提出估價單的期限。

第一段的 Please be informed that... 是在傳遞壞消息時經常使用的刻板形式。

第二段的 Please cancel the description， cancel 是我方的動作，而非估價者的動作。另外，最好別用 cancel 這種極度否定的字眼。

第三段中的 We regret to inform you... 表達遺憾之意，但是最好想想其它表達方式。

接下來看看改善範例。改善範例是從「總論到細節」，而且相較於原文以壞消息（不使用單位式空調系統）為開頭，改善範例則以好消息（使用中央空調系統）做起頭。

例文 10 的改善範例（○）

① As a result of a revision by the client[1], the following change has been made: central air conditioning will be employed instead of unit air conditioning. As a result, the following changes have been made to the scope of supply[2]:

② Added to the scope of supply: central air conditioning system[3] specified in the accompanying sheet.

Removed from the scope of supply: unit air conditioners[4] described in 2.2 of our Specification.

③ In spite of this change, there has been no change in Client's construction schedule, so may we ask your cooperation in submitting your quotation, reflecting this revision, by May 10 as in our original Request for Quotation?[5]

④ If you have any questions regarding the details of this change, please contact me at any time.

1) client 客戶（請看參考部分） 2) scope of supply 供貨範圍
3) central air-conditioning system 中央空調系統 4) unit air conditioner 單位式空調機 5) request for quotation 委託估價（單），縮寫為 RFQ

【段落大綱】

① 傳達計畫變更的概要。
② 變更的結果，追加的供貨範圍和刪除的供貨範圍。
③ 再次確認估價期限。
④ 如有問題請告知，並作結。

【內容】

由於客戶重新評估的結果，產生下列變更。換言之，將採用中央空調，而非單位式空調。因此採購範圍也有下列變動。

追加項目：附件所示中央空調系統

從供貨範圍中刪除者：本公司規格書 2.2 項所示之單位式空調機

縱有此變更，客戶的建築時間表並沒有變更，所以請協助於當初委託估價的行程 5 月 10 日以前，提出反應上述變更的估價單。

針對此一變更，如有疑問，請隨時與我方連絡。

（參考）「顧客」是 customer 或 client

在字典裡，「顧客」的解釋為 "a customer; a patron; a buyer; a client（法律上的）"。這裡所謂「法律上的」，是指律師的委託人。商店或廠商的顧客，一般而言是 customer，看看各公司的簡介或書信，常常以 customer 來稱呼客戶。但是 client 也可以作 customer 的用法。

接下來，我們看看 *LDCE* 的 customer 解釋，依 USAGE 區分如下：

*When people go out to buy things in shops, they are **shoppers**: a busy street full of **shoppers**. When people buy from a particular shop, they are that shop's **customers**: Mrs Low can't come to the telephone— she's serving a **customer**. If you are paying for professional services, e.g. from a lawyer or a bank, you are a **client**. If you are staying in a hotel, you are a **guest**.*

LDBE 的 client 解釋為下：

(1) *a person who employs or uses the services of a professional*

adviser (other than a doctor or dentist) esp. of a lawyer, architect,
stockbroker.　⑵ a customer, a person who buys or is likely to
buy from a seller.

⑴中的 architect（建築師），可以證明這封建築師事務所所
寫的例文 10, client 會比 customer 更合適。而⑵的說明中也提
到，client 有時候的用法相同於 customer。

我們來看看美國 *Quality Progress* 雜誌 1992 年 5 月的報導，
觀察 customer 和 client 的用法。

*Leaders spend lots of time with **customers**, but they don't just*
*spend with their executive partners.　They get down into **client***
organizations and meet with the people who are actually using
their products and services.

（廠商的領導人花很多時間跟他們的 customers 一起，而不
僅是和高階同事在一起。領導人深入 client 的組織，和這些
實際使用他們產品或服務的人見面。）

這裡的 client 就和 customer 的用法相同。但是有些人會認
為 client 比 customer 正式，階層也比較高。

4. 認可或建議

發包廠商認可承包商所提出的技術規格是有責任和風險的，
所以經常附帶許多條件，或是頻頻要求對方提供資料，事事謹
慎。相對地，承包商方面則是要獲得發包廠商的認可，之後才可
安心地進行，所以也想盡可能地獲得多項認可。

至於那些不需要發包廠商一一認可的事項，就必須由發包廠
商通知承包單位，請他們自行判斷。不過在技術方面，發包廠商
應該毫不保留地給承包單位建議。

例文 11是發包公司 ABC 對承包商 XYZ 公司的回應，XYZ

公司設計 XXX 運輸用貨櫃，想請 ABC 公司認可。

例文 11　貨櫃規格的建議（原文）

Re:　XYZ-Designed XXX Shipping Container

This is in response to your letter of September 5th.

ABC is not in a position of approving the XYZ-designed XXX shipping container, but provides consultation.

According to your letter, ABC understands that XYZ's XXX container meets the specification of ABC's domestic container in terms of shock and insulation resistance assuming that XYZ uses the chemically treated, anti-static version of foam.

The container needs to be transported and handled in such a manner that the maximum shock (up to 20G) can be guaranteed for XXX inside the container in transportation and handling.

XYZ may use the container in your normal worldwide distribution operation, considering the above.

For our evaluation purpose, please send us the sample of the foam.

Sorry for our delayed response on this matter.

Thank you for your co-operation.

Best regards,

【內容】

　　主旨：　XYZ 公司設計的 XXX運輸用貨櫃

　　本函乃在回覆您 9 月 5 日的來信。

　　本公司並無立場認可貴公司所設計的 XXX 運輸用貨櫃，但會提供建議。

　　根據您的來信，本公司了解貴公司的 XXX 貨櫃使用經過化學處理的抗靜電發泡體，在耐撞擊性和絕緣抗阻上，符合本公司國內貨櫃的規格。

　　運輸和作業中的貨櫃，必須保證最大 20G 的撞擊值以保護貨櫃內的 XXX。

　　考慮上述原因，此貨櫃適用於一般全球運輸作業。

　　本公司為了評估，請寄來發泡體的樣本。

　　關於本件的回覆延誤，請見諒。

　　感謝您的合作。

【問題點】

　　這篇例文的問題點，在於表示自己 ABC 公司沒有立場認可XYZ 公司設計的貨櫃，卻又煞有介事地要求寄送資料以供研究。到底是要認可或只是單純的建議，實在模糊不清，甚至會造成誤會或糾紛。 XXX 是裝在此貨櫃中的產品名稱。

　　第二段 ABC is not in a position of approving...的語意，在查過 *LDCE* 的 position 後，有 I'd like to help you, but I'm not in a position 的例句（=I can't）。也就是說，不單純是 I can't，另外還有「想幫忙但沒辦法」的語意，也就是不得不的意思。這

種迂迴的說法，主要用在無法依照對方的願望，或是婉拒對方要求的時候。至於這篇例文中， not in a position 的語意如果是 ABC 公司想要認可，但基於某些原因無法做到，但又不說明是什麼原因的話，那麼 XYZ 公司豈不手足無措。換句話說，到底是不合格，還是資料不足無從認可，抑或是並非 ABC 公司的認可範圍，可由 XYZ 公司自行判斷處理，總之要給予對方明確的答覆。

consultation是「諮詢、商量」的意思，所以 provide consultation 在此並不適用。到底是要給建議或是提意見，應該像改善範例一樣，明確表示是要「提出評語」。

第三段 According to your letter, ABC understands...「根據您的來信」，彷彿是在懷疑對方。但是，由於必須要有「從您提供的訊息判斷」的前提，所以最好用改善範例的表達方式。

in terms of ～是「就～而言」。

第四段 The container needs to be transported and handled...，用 the container needs to be (must be, should be) used 就可以。

the maximum shock can be guaranteed for XXX 這句話的意思是「保證對 XXX 有最大衝擊值」，而這裡想說的是「衝擊最大也不能超過 20G」，所以意思剛好相反。另外 maximum shock 加上定冠詞 the表示彼此都知道最大衝擊為 20G，但實際上這裡的情況並非如此。 G是加速度的單位， $1G=9.80665m/s^2$。一般人體的限度是 5–6G（《單位辭典》）。另外，國際單位 SI 中明定加速度單位為 m/s^2，同時也認可 Gal (Galileo, $1Gal=1cm/s^2$) 的使用。

第五段 XYZ may use the container, 一開始說 ABC 公司沒有立場認可，但這裡卻又認可貨櫃的使用，讓人不清楚 ABC 公司到底想做什麼。

第六段要求寄送樣本 For our evaluation purpose, please send

us... 這裡的 for our evaluation purpose 指的是為了認可而需要評估的意思嗎?

最後的 Thank you for your co-operation. 是感謝對方寄送樣本嗎? 沒有特殊用意的話, 用不著道謝。

例文 11 的改善範例 (○)

Re:Evaluation of XXX Shipping Container Designed by XYZ

① In answer to your question in your letter of September 5, ABC's approval for this container is not required. We are, however, willing to offer our evaluation in response to your request.

② Our evaluation of this container, based on the information you supplied, is that it meets the specifications for shock resistance[1] and heat resistance[2] for the container used by ABC in the domestic market. This assumes that it is made from chemically treated, anti-static foam[3]. If it can be guaranteed that shock during transportation will not exceed 20G, this container can be used in normal worldwide distribution[4].

③ If you would like a more detailed evaluation, please send us a sample of the foam.

④ Sorry for the delay in this response.

Best regards,

---

1) shock resistance　耐撞擊性　2) heat resistance　耐熱性　3) anti-static foam　抗靜電發泡體　4) worldwide distribution　全球運輸

【段落大綱】

① 說明本傳真的前提:「沒有認可貨櫃的必要,但是要評估」。
② 給予「可使用,但有附帶條件」的評價。
③ 再指示接受更進一步評估的順序。
④ 對延遲回覆表示抱歉,並作結。

【內容】

主旨: XYZ 公司設計的 XXX 運輸用貨櫃評估

回覆您 9 月5 日的來信,此貨櫃不須 ABC 的認可。但是,針對您的要求,本公司可提供評估。

根據您的資料,本公司對此貨櫃的評估為,符合本 ABC 公司國內市場使用貨櫃耐撞擊性及耐熱性規格,這是因為我們設定此貨櫃乃以經過化學處理的抗靜電發泡體製作。如果能夠保證運輸期間的衝擊值不超過 20G,此貨櫃將可用於全球的一般運輸。

如果需要更詳盡的評估,請將發泡體的樣本寄來。

很抱歉延遲回覆。

【注意事項】

改善範例的立場是認為該公司沒有認可貨櫃的必要,但是給予 XYZ 公司的貨櫃符合該公司的國內規格,以及若符合特定條件並可全球使用的評價。此外還提出如果送來樣本,將更詳盡地加以評估。

5.　錯不在己, 就不用認錯

　　我們習慣錯不在己, 也向對方道歉, 這種習慣對歐美人而言, 非常不自然, 等於是沒必要, 卻讓自己處於不利的立場。這種表達方式很可能就在一個動詞中顯現出來, 要特別小心。

例文 12　討論提升生產線產能 (原文)

Regarding your FAX 6–3–120:

1. Basically we admit we have to meet 200,000 pieces/year.

2. But in order to modify model AAA line to model BBB line, we have to finish 19xx MY (Model Year) model AAA production before 19xx April. So in this case 200,000 pcs can't be produced as 19xx MY total. (This is not official.)

3. Because of the situation of the above 1 and 2, we need 1 or 2 days to study how we can meet 200,000 pcs/year.

So, we are really sorry, but please wait for our official reply by 6/7.

【內容】

　　關於您 6–3–120 的傳真:

　　1.基本上, 我們承認必須達到每年 20 萬個。

　　2.但是, 如果要把AAA 型生產線改為 BBB 型生產線, 就必

須在19xx 年 4 月以前結束 19xx 模型年的 AAA 型生產。
如此一來, 19xx 模型年總生產額將不到 20 萬個（這並
非正式回答）。

3.由於上述 1、 2 的情況, 我們需要一、兩天研究如何達到
20 萬個的條件。

很抱歉, 正式回答請等到 6 月 7 日。

【問題點】

第一段 Basically we admit（基本上我們承認）含有「不承
認」的意思, 而且 to admit 的意思是「承認事實」, 通常用於
She admitted stealing the bicycle.（她承認偷了腳踏車）這種做了
壞事或犯錯的時候。如果用在例文中, 會給人由於自己的疏忽才
會造成無法達到 20 萬個的印象。如果自己沒有過失, 在這裡最
好不要用 admit 這個動詞。

第二項以及最後的 official reply 表示這還不是正式回答, 正
式回答要等到 6 月 7 日。區分正式和非正式, 會讓人覺得這樣
的做法太狡猾, 要特別小心。

最後的 we are really sorry 真的有必要說嗎? 而且, 也不是
please wait for our official reply by....,應該改為 please wait for
our reply until....（在～之前, 請等候我方回答）, 或是像改善範
例的 we will reply by....。另外6/7的日期寫法未必代表 6 月 7 日,
也可以解釋成 7 月 6 日, 所以最好寫成 June 7 或是 7 June。第
二項19xx April 的正確寫法是April 19xx。

例文 12 的改善範例（○）

We understand the need for 200,000 pieces/year. We will
have some difficulty, though, in increasing 19xx model year

> (MY) production to this amount, since the whole 19xx MY model AAA production must be completed before April 1, 19xx. We are now studying how this need can be met, and will reply by June 7.

【內容】

我們了解每年要生產 20 萬個。但是，19xx 模型年的 AAA 型生產必須在 19xx 年 4 月 1 日前結束，所以很難將 19xx 模型年的產量提升到這個數量。如何才能達成，現在正在研究之中，我們會在 6 月 7 日前回覆。

【注意事項】

因為是傳真，所以只寫要點。比起原文的 So in this case 200,000 pcs can't be produced...，改善範例的 We will have some difficulty...來得委婉多了。

6.　體貼顧客

近年來 customer satisfaction（顧客滿意）越來越受到重視，就這一點而言，一般人都認為我國企業對客戶的服務非常良好、非常重視客戶。的確，企業之間競爭激烈，重視市場的我國企業確實要比歐美公司體貼客戶。

下面這封信，是 ABC 公司的工程師對出口到海外的該公司電腦損壞情況表示看法。不忘維護公司利益，遵照商務倫理來闡述見解是件好事；但是這封信如果給人想要開脫，或是讓客戶覺得這家公司根本撒手不管的話，就說不上重視顧客了。

例文 13 處理電腦損壞意外 (原文)

Subject: Repair on the damaged C60 CPU

We notice that we CANNOT repair the following C60 CPU which was damaged during transportation:

Model No.   A–C60–32KV CPU
Serial No.   1907
FOB date 10 September 199x

Judging from the photographs of the inside and outside C60 CPU (Central Processing Unit), it is no doubt that a big shock in excess of the maximum tolerable level was given to it during transportation. Due to the shock many parts, such as VLSIs, printed-circuit boards and pin connectors, must have been damaged seriously. This is why we concluded that we cannot repair this C60 CPU.

Yours sincerely,

【內容】

　　主旨: 受損 C60 型 CPU 的修理

　　本公司無法修理運輸途中損壞的下述 C60 型 CPU。

　　　　型號:　　　A–C60–32KV CPU

　　　　生產號碼: 1907

　　　　裝船日期: 199x 年 9 月 10 日

從 C60 型 CPU （中央處理器）的內、外部照片看來，可能是在運送途中遭到最大承受度以上的撞擊。因為這種撞擊，使得 VLSI、列印電路板、外部接頭等多數零件受到嚴重損傷。這是本公司判斷無法修理 C60 型 CPU 的原因。

【問題點】

這封信的文法錯誤很少，內容也容易了解，但是語氣上似乎有棄客戶於不顧的感覺。

第一段 to notice 表示 to pay attention to 或「注意」的意思，理論上這種用法並不正確。接下來， we CANNOT 以大寫來強調，這種否定的表達方式帶有攻擊性，對客戶而言，在態度上有問題。

最後一段的 VLSI (very-large-scale-integrated circuit) 是簡寫，但 printed-circuit boards 卻沒有簡寫， CPU 又有說明，像這樣的用法實在不值得學習。一般人可能知道 CPU 這個簡寫的意義，但或許不知道 VLSI （超級 LSI＝超大型積體電路）。即使對方懂得 VLSI，但是難懂的單字用簡寫，簡單的單字卻用簡寫加上說明，用法上稱不上統一。請參考後面「縮語、簡稱的規則」。

must have been damaged 是指「從理論上判斷，應該損壞了」？或是觀察後推定的原因呢？這裡的說明不清楚。

we concluded that we cannot repair...是最後的答覆，但這只是將客戶推開、置之不理的做法。讀者看這封信時，一定覺得不知所措，進而覺得不滿。

例文 13 的改善範例（○）

Subject: Extent of Damage to C60 CPU

Model No.　　A–C60–32KV　CPU
Serial No.　　1907
FOB date 10　September 199x

　Upon inspection, we found that this CPU has received extensive damage to its components, including VLSIs, printed-circuit boards, and pin connectors. We regret to report that damage of this extent cannot be repaired.

　CPUs are packed for shipping to protect them from any reasonable impact during transportation. The extent of the damage to this CPU leads us to believe that it must have received a shock considerably more severe than in normal circumstances.

【內容】

　　主旨：　C60 型 CPU 的損壞程度
　　　　型號：　　　A–C60–32KV　CPU
　　　　生產號碼：　1907
　　　　裝船日期：　199x 年 9 月 10 日
　　檢查的結果，得知此 CPU，包括 VLSI、印刷電路板、外部接頭等眾多零件均遭損壞。很遺憾，這種程度的損壞已經無法修理。

　　CPU 出貨用的捆包是用來保護運輸途中一般的衝擊。此 CPU 受損的程度讓我們相信，應該是受到比一般狀況還要嚴重的撞擊所致。

【注意事項】

　　在寄發信函之後，必須委託運輸公司或保險公司採取適當措施，視實際情況處理。

（參考）縮語、簡稱的規則

　　我們經常使用英文的縮寫，例如高科技領域中的 AI、CAD、CAM、CIM、FA、RAM 等，這些縮寫一般都認得，至於全名可就不曉得了。在我們所寫的信函之中，經常突然用縮寫，這等於是忽視了讀者。歐美人在撰寫文獻或論文時，極少突然間就使用縮寫的情況。縮寫還是有一定原則的，接下來將 *Chicago Manual* 及其它參考書中的規則整理如下：

　　1.使用縮寫之前，先拼出原本的單字，在括弧內寫上縮寫，之後才只用縮寫。例如 artificial intelligence (AI); flexible manufacturing system (FMS); Japanese Industrial Standard (JIS)。而文章中如果不使用這些縮寫，就不用標示縮寫。

　　2.文章標題不會使用不為人知的縮寫或簡稱。在節錄中最好也不要用縮寫或簡稱。

　　3.表示單位的單字，前面如果沒有數字的話，便不能省略。例如 The measurement is converted into cubic centimeters.（不簡寫為 cm^3）。

　　4.如果認為讀者能夠了解，專業術語有時也可予以簡寫。例如 AC 或 a-c (alternating current); C (centigrade)。

5.複數形：

(1)縮寫的複數，往往直接加 s 不用一撇。例如two CPUs。

如果縮寫的字尾是 s，就加上's。例如 10 BPS's。

(2)單位的縮寫通常作單數。例如 kg（不是 kgs）。

6.縮寫後加不加句點，要看習慣。例如 a.m. (AM); c.o.d. (COD); f.o.b. (fob,FOB); a.c. (a-c); Fig.（不是 Fig）。

單位的縮寫不加句點。例如 2 km; 5 ft 7 in.（inch的縮寫為了要與介系詞區別，所以不是 in 而是 in.）。

7.文中用記號不如用縮寫。例如 8 in.（不是 8"）；12 by 15 ft（不是 12'× 15'）；percent（%比較不好）。

8.縮寫以小寫為原則（ SI unit（國際單位）以小寫居多）。正式名稱如果是以大寫開頭，則簡寫時也大寫。

不過，這些原則也沒有被嚴格遵守。像第 7 項中提到文中必須寫成 percent，但是實際書信中還是有使用%的情況。

7.　用肯定式而非否定式

經常會有人告訴我們，用肯定式而非否定式。例如要告知打烊的時間：

Closed Saturday at noon　　　　　　　　（否定的）

Open until noon on Saturday　　　　　　（肯定的）

因為肯定的表達方式給人積極的印象。我們來看看下面這份傳真的例子。

例文 14 預告出貨（原文）

DATE: 7/17/199x   Reference No. A7–17–05

TO: XYZ Mr. R. Doe   From: T. Henry

CC: ABC Mr. U. Smith   CC: Y. Johnson

SUBJECT: Filters for Sample

Order No.	Flight date	Q'ty (pcs.)
ABC012345	8/28	10

The above timing is not official answer from the prototype shop. But it is estimated by them. I negotiated with them to send these samples before our summer vacation (8/10–8/18). But they said that they can't make them in such a short time. So I made them estimate the timing which is above figure.

I'll tell you the official answer (timing) as soon as I get the information.

                 Best regards,

【內容】

    主旨：樣本用濾網

接單號碼	班機日期	數量（個）
ABC012345	8/28	10

上述時間並非製模工廠的正式回答，只是預估額。原先交涉在本公司暑假 (8/10— 8/18) 之前寄出樣本，但是他們說無法在這麼短的期限之內完成。因此，經他們預估的交貨期限如上。

一有新的消息，我會正式地告知您（時間）。

【問題點】

「非正式回答」、「這麼短的時間做不出來」等否定的表達方式不斷重複，讓人覺得這消息實在不可信賴。另外，official answer 也是不斷重複，讓人感覺好像另有弦外之音。

這封短信到底在說什麼，實在是交代不清。原因是沒有在 Subject 說清楚主題，而 Subject 之後竟然立刻就是一個表。在 timing 上也很模糊。如果先有 Subject，之後再標示 delivery date（樣品交貨期）的話，讀者很快就能掌握內容。

例文 14 的改善範例 (○)

Subject: Sample Filters: Delivery Date

① The prototype shop estimates that the sample filters will be shipped as follows:

    Order No. : ABC012345
    Flight Date: August 28
    Quantity   : 10 pcs.

② Even this date is tentative[1]. I will let you know a firm date[2] as soon as I have it, by August 23 at the latest.

1) tentative　暫定的　　2) firm date　確定日期

【段落大綱】

① 回答可出貨的數量與交貨期。
② 先説明上述日期為暫定，並告知哪一天將正式回覆確定交貨日。

【內容】

　　主旨：樣本用濾網：交貨期
　　製模工廠預估上述樣本將如下列日期出貨。
　　訂單號碼：　ABC012345
　　班機日期：　8 月 28 日
　　數量　　：　10 個
　　此日期乃暫定，當得知確定日期後，最晚會在 8 月 23 日前通知。

8.　說出自己的意見

　　我們經常不說自己想説的話，而是借用他人的話來跟外國交涉。舉例而言，關於設計、生產和交貨期等等，公司會利用生產廠商的藉口；而生產廠商則是在技術或基本設計上借用授權者的說詞。不過，今日我國技術已經達到國際水準了，生產廠商即使不靠公司同樣也能跟海外溝通，所以實在不需要再借用別人的意見，應該從本身立場清楚闡述看法，這樣才能讓對方充分認同我方的正當性。

　　依賴他人的意見，就等於承認自己的能力和知識不足，有時還會被認為是轉嫁責任。即使在引用他人（第三者）的見解時，只要這第三者就溝通對象而言視同為我方人員（同一家公司），那也該當作是自己的見解，傳遞給對方。

例文 15 拒絕分售零件的傳真 (原文)

XYZ Electric Ltd.   May 27, 199x

To   : Tony Doe   Total page 1

From: H. Benjamin (Sec. 3)

cc   : K. Arai
Re   : Your fax dated May 19

I got your recent fax this morning for the first time. I guess
it was missing. Anyway the following is the answer for you.

1. Component XXX for ABC Computer

Regarding your inquiry of PCB-excluded Component XXX, I
checked with Taipei Plant. They said the product has an LED
sensor which are mounted in the PCB. It is Taipei's custom
part. So it is not possible to sell Component XXX separately
from the PCB. Even a case the you/customer buy the sensor
from Taiwan, from the engineering point of view, you/customer
will face a difficulty to adjust the sensor's location on the PCB
for the specified reliability. So there are two reasons for ABC
not to sell Component XXX separately from PCB; company
policy and technical difficulty.

:

【內容】

　　今天早上才收到您最近的傳真。可能是弄丟了。無論如何，以下是給您的回覆。

　　1. ABC 電腦用零件 XXX

　　關於您洽詢不包含 PCB 的零件XXX，我跟臺北廠確認過了。根據臺北廠的說法，本產品是將 LED 感應器載於PCB 中。這是臺北廠的特製零件。因此，零件 XXX 無法與PCB 分開銷售。即使貴公司或顧客另從臺灣採購這個感應器，從工程的觀點來看，貴公司或顧客恐怕很難藉由調整 PCB 上的感應器位置，達成指定的信賴度。因此，本公司不將 PCB 與XXX 零件分開銷售的原因有二，乃是因為公司的方針和技術困難度的關係。

【問題點】

　　傳真開頭的 cc: 表示副本寄予第三者，但是這裡到底指的是發信者傳來的，或是請求轉給收信者的，沒有交代清楚。

　　導入正文的部分在辯解為什麼傳真會延誤，但是 I guess it was missing. 則是多餘的。下面的 Anyway 常常用來表示「總而言之」，但在英文中是很沒有禮貌的用法。

　　本文第二句 which前面的 an LED sensor 是單數，但是卻用which are mounted 接複數，這是文法的錯誤。此外，LED = light emitting diode（發光二極體），而 PCB 是printed circuit board（印刷電路板）的縮寫，這些縮寫如果當事人都很清楚的話，就不需要全部拼出來（參照第 6 節「縮語、簡稱的規則」）。

　　本文第五句最開始的 Even a case 應該是 Even in case。the you/customer 則應該是 you/ the customer 才正確。但仍請鎖定 you 或 the customer 其中一個就好。而這裡的「即使貴公司（或顧客）另從臺灣購買」，其實很有挑釁的味道，請參照改善範例。

最後一句 there are two reasons for ABC not to sell Component XXX separately from PCB; company policy and technical difficulty. 在列舉的時候, 不應該用分號 (;), 而是用冒號 (:)。

這篇例文一直在重複理由, 沒有告訴顧客賣方希望他們怎麼做, 會給對方棄我於不顧的感覺。

例文 15 的改善範例 (○)

1. Component XXX for ABC Computer

① Regarding your inquiry for a PCB-excluded Component XXX, we are unable to supply it separate from the PCB. The PCB contains an LED sensor which is essential for its operation. If the two are purchased separately, it is very difficult for the user to adjust the sensor's location on the PCB to achieve specified reliability.

② We are sure that the PCB-equipped Component XXX is the most suitable for the ABC computer, and would be pleased for the chance to supply it to you.

【段落大綱】

① 說明印刷電路板無法與零件 XXX 分開銷售以及其原因。
② 強調不要與印刷電路板分開使用才是最佳狀況。

【內容】

　　關於貴公司洽詢不包含 PCB的零件 XXX，但是此零件無法與 PCB 分開銷售。PCB 包含啟動時不可或缺的 LED 感應器。如果將這兩個零件分開購買，使用者很難調整 PCB 上的感應器位置，以獲得指定的信賴度。

　　我們深信載有 PCB 的XXX 零件最適合 ABC 型電腦，希望有機會能將此產品供應給貴公司。

【注意事項】

　　改善範例中沒有提到臺北廠的事情，全都以自己的立場來說明，這是一般對客戶的因應之道。如果對方是同一公司內部的人，而且又知道發信者的技術知識不足信賴時，一定要明示消息來源。

　　另外，「即使貴公司或顧客另從～購買」和「誰～」，這種表達方式由於給人挑釁的意味，所以改善範例中改以被動式表達。而後一句的動作主詞既不是 you，也不是the customer，而是 the user。

　　最後是賣方對買方的請求。

第IV章

請求

1. 考慮對方的立場和情況

　　有很多業務員經常使用「我剛好到這附近」的藉口，在沒有事前預約的情況下，就去拜訪客戶，這跟歐美的習慣相違背。在歐美，拜訪某人之前一定要事先預約，這是一種常識。預約時，就要把來由先說清楚，同時還得考慮對方的國家、地方或企業在你希望拜訪的那一天到底是什麼狀況。可能因為你在暑假時不經意的拜訪，使對方的公司以為有什麼大事，還把休假的負責經理從渡假地叫回來呢！

　　下面這個例子是工程師青木先生給美國某公司的傳真，表示希望前去拜訪。

例文 16　請准拜訪的傳真（原文）

> Feb. 19, 199x
>
> Ms. Linda Wilson
> XYZ Laboratories Inc.
> 1655 Scott Blvd.
> Santa Clara, CA 98050
>
>
> Subject: The ISO 9000 Registration Program
>
>
> Dear Ms. Wilson:
>
>
> We are very much interested in your program mentioned
> above.

I'm a quality staff engineer of ABC Electric in Japan. (ABC is one of the biggest electronics parts maker in Japan.) We have 10 divisions in Japan and also many plants in all over the world, including the factory in Oakland, CA.

I'm planning to visit USA to investigate and study the situation of ISO 9000 from Feb. 23 to Mar. 4.

I would like to visit XYZ around noon on Feb. 24.

It would be very appreciated if you could arrange the meeting and let me know to whom and when can I meet, as soon as possible.

Yours sincerely

Kazuo Aoki
Quality Management Division
ABC Electric Co., Ltd.

Phone　+81–3–xxxx–xxxx
Fax　　+81–3–xxxx–xxxx

【內容】

　　主旨: ISO 9000 登記計畫

　　我們對於貴公司的計畫非常有興趣。

　　我是日本 ABC 電機的品質工程師。（ABC 公司是日本最大

的電子零件廠之一。) 本公司在日本有十個部門，在全球各地也有許多工廠，其中之一就在加州的奧克蘭。

為了調查和研究 ISO 9000，我預定於 2 月 23 日到 3 月 4 日之間訪問美國。

我希望在 2 月 24 日中午拜訪貴公司。

請代為安排會議，並儘早通知我該在什麼時候見什麼人。

【問題點】

傳真日期是 2 月 19 日，希望拜訪的日期是 2 月 24 日，等於回覆的傳真必須在 2 月 21 日以前傳到青木先生的辦公室，如果從加州就得在 20 日傳真。所以負責回覆的琳達就只有 19 日和 20 日兩天準備。這麼緊迫的時間，根本沒辦法向青木先生詢問想見什麼樣職務的人，或是因為時間無法安排，請求改換日期了。

青木先生想在中午拜訪 XYZ 公司，但是中午去拜訪別人，對琳達小姐及其它人而言，都是很令人厭惡的。

本文第二段第二句 in all over the world 的 in 應刪除。同一句的 the factory in Oakland 等於是「(你也知道)奧克蘭工廠」，但實際上可能沒那麼有名，所以冠詞 the 應該改為 a。

這段文字中，介紹公司時出現了三次 in Japan，實在很礙眼。公司介紹最好另附一張說明。

最後一句 It would be very appreciated if you could arrange the meeting and let me know to whom and when can I meet...看起來好像很有禮貌，其實是沒有給對方說不的機會，是 top-down (上對下) 的表現。此外，不是 when can I meet 而是 when I can meet。拜託別人代為安排會議時，一定要告知會議的內容是什麼，希望向誰請教。

例文 16 的改善範例（○）

Subject: Study by ABC Electric Co. of ISO 9000 Registration Program

① We have read materials on your ISO 9000 Registration Program, and we are very interested in learning more about it. We in particular hope to be able to study:

  1) Registration procedure
  2) Technical requirements

② I will be visiting Santa Clara on February 24. I know this is very short notice[1], but I wonder if it would be possible to visit your office early that afternoon[2]? If you could arrange for someone to speak[3] with me about these items I would be very grateful[4].

③ For your reference, I am also sending information about ABC Electric with this fax. I look forward to hearing from you.

1) short notice　突然的通知　　2) early that afternoon　當天下午最早的時段，若單指「下午最早的時段」，則是 early in the afternoon
3) to arrange for someone to speak　安排某人說話（=to arrange that someone speaks）　　4) be grateful=be feeling thanks

【段落大綱】

① 表示對對方的計畫有興趣，並說明希望了解的項目。
② 說明希望拜訪的日期，請求適當人選來作說明。
③ 以表示同時附上自己公司的資料作結。

【內容】

拜讀貴公司 ISO 9000 登記計畫的資料後，我們非常有興趣，希望能夠更深入了解，尤其想要學習：

(1)登記程序
(2)技術條件

我將於 2 月24 日訪問聖克拉拉。我知道這個通知非常急促，但是希望能在當天下午最早的時段拜訪貴公司。若您能為我安排人員會面，我將十分感激。

茲將 ABC 公司的相關資料隨此份傳真附上，供您參考。期待您的回答。

【注意事項】

這種突如其來的要求，最好用帶有躊躇口吻的 I wonder if it would be possible...，而不是原文的 It would be very appreciated if...。關於這種類型的表達方式，英國的技術寫作指導師 Kirkman (1992) 就曾寫過下述的評論：

I have been told firmly that the British English preface to a request or instruction "I wonder if you would mind..." can seem pretentious or stuffy to Americans.

（曾經有人斷然告訴我，英式英文的委託或指示的前置說詞 I wonder if you would mind... 對美國人而言，是既做作又

嚴肅的。）

英國人的英文的確不同於美國人的英文，我們除了學習英式和美式的差異之外，最好還能學習有禮貌的英文。即使禮貌過頭，也不會有人見怪。此外，美國人也會用 I wonder if you would mind... 的說法。

本文第二段 I will be visiting Santa Clara on February 24. 為什麼要用未來進行式呢？從我們的觀念來講，根本無從區別與 I will visit Santa Clara on February 24.（我將於 2 月 24 日訪問聖克拉拉）的差異。I will visit Santa Clara 是包含意圖在內的未來，而 I will be visiting Santa Clara 則不包含意圖在內，單純只說行程上會是如此。所以在表示拜訪行程時，大多會使用未來進行式。

當然也可以用現在進行式 I am visiting Santa Clara on February 24. 這也是在說明未來預定的行動。

2. 不是為了自己，而應替對方考慮

運送貨物時，最好避免載運會爆炸或起火、有腐蝕性、對人體有害的危險物品。這種危險物品最好在當地採購。但如果是只想省去自己在出口手續上的繁文縟節這個原因，反而會給人自私自利的感覺，必須站在對方的立場來說明理由。

例文 17　委託對方於當地調查危險物品的採購事宜（原文）

Re: DANGEROUS MATERIALS

Please refer to the following list of materials we supply. You would find dangerous materials in the list.

Our procedure of export dangerous materials is so complicated that we would like to avoid doing these procedures, otherwise would like to minimize dangerous items then standardize those procedure.

So please advise us whether or not you are able to buy those dangerous materials at your side, and which dangerous materials we must supply.

For your information, a definition of Dangerous Goods is as follows:

Dangerous Goods

1. Explosive
2. Gases, compressed, liquefied, dissolved under pressure or deeply refrigerated
3. Flammable liquids
4. Flammable solids
   Substances liable to spontaneous combustion
   Substances that, on contact with water, emit flammable gases
5. Oxidizers
   Organic peroxides
6. Poisonous substances
   Infectious substances
7. Radioactive substances
8. Corrosives

9. Miscellaneous dangerous goods
*Details of regulation depends on material's spec. and methods of transportation.

Above items are dangerous goods which must be transported under UN regulations.

Awaiting for your reply.

【問題點】

　　這裡我們想要鎖定的問題點，在於第二段的「出口危險物品的手續非常複雜，最好能夠避免。」(we would like to avoid doing these procedures)因為手續繁複，所以希望能省即省，這都是只考慮自己的立場，沒有替對方著想。更何況，國際機構或國家制定的手續過度繁縟所以不想做，說這種話本身就有問題。即使心裡真的是這麼想，也應當盡量避免這種發言。

　　標題中只提到「危險物品」，到底是什麼卻沒有交代清楚。如果寫成「委託於當地採購危險物品」的話，閱讀者可以一開始就有心理準備。

　　本文第一段 Please refer to the following list of materials we supply. （請參照我們所供應的下列材料清單），而底下則是列出一般危險物品種類的清單。「我們所供應的材料清單」到底在哪裡？

　　第二段 otherwise 的用法錯誤。如果想說「希望避免複雜的手續，換句話說，希望將危險物品的量降到最低，使之合乎標準」，那可以分成二個句子來寫：We would like to avoid doing these procedures. We would like to minimize dangerous items transported, then standardize..., 而不要用 otherwise。

　　第四段的 a definition of Dangerous Goods is as follows,
底下只列舉危險物品的種類,卻沒有 definition (定義)。而 a
definition 表示眾多定義中的一種,這裡如果要表示特定的定義,
該用 the definition 才對。

　　表的結尾打上 * 號 (asterisk),這時候必須標示它指的是本文
的哪一部分的註解。也就是說,本文的那個部分也要打上 * 號。

　　最後 Awaiting for your reply. 的 await 是及物動詞,所以不
需要 for; 而使用不及物動詞 wait, 則應該寫成 Waiting for your
reply. 無論哪一種寫法,這種草率的結尾都不好。

例文 17 的改善範例 (○)

Re: Request for Local Sourcing[1] of Dangerous Materials

① We would like to request that dangerous materials for
production be obtained locally whenever possible. Therefore may we ask you to check the list accompanying[2], and
let us know which of these materials are available to you
locally?

② The procedures for export/import of these materials are
very complicated and time-consuming[3]. In order to avoid
unnecessary trouble at your end and ours, and the pos-
sibility of unforeseen[4] production interruptions[5] from
government inspection delays, we would like to determine
which items can be obtained locally and which not, and
then standardize procedures for those we must continue

---

to supply.

③ For your information, the following classes of materials are defined by UN regulations[6] as "Dangerous Goods[7]".

1. Explosives
2. Gases, ...
3.
   :
9. Miscellaneous[8] dangerous goods

④ We would greatly appreciate your reply by March 22, so that we can...

Regards,

---

1) local sourcing　尋找當地的供應商，如果不想用 source 的動詞形，也可以寫成 Request for Surveying Local Availability of Dangerous Materials　2) the list accompanying　附件清單　3) time-consuming　耗時的　4) unforeseen　無法預測的　5) production interruption　中斷生產　6) UN regulations　聯合國規定　7) dangerous goods　危險物品　8) miscellaneous　各式各樣的

【段落大綱】

① 說明希望在當地採購危險物品的前提，之後請求代為調查於當地採購的可能性。
② 說明於當地採購的理由，以及希望將當地採購項目標準化的意向。

③ 提示危險物品分類。
④ 告知回答的日期（及其理由）。

【內容】

主旨：委託調查當地購買危險物品的可能性

我們希望生產用的危險物品能夠儘可能地在當地購買。所以煩請確認附件清單，看看這些材料中哪些可以在當地採購，並告知結果。

這些材料的進出口手續非常繁雜，而且曠日耗時。為了避免雙方無謂的麻煩，以及因官方檢查延遲導致無法預期的生產中斷，請決定哪些項目可於當地購得，而哪些無法於當地購得，我們想將必須繼續供貨的項目順序標準化。

依據聯合國規定，下列材料的種類定為「危險物品」，請參考。

1.火藥類
2.瓦斯
3.
    ：
9.各式危險物品
希望能在 3 月 22 日以前獲得答案，好讓我們能夠……

【注意事項】

首先請注意注意段落的順序與原文不同。原文是⑴希望對方看附件的清單；⑵出口手續繁雜，所以想減少出口項目，並將之標準化；⑶所以希望對方告知在當地能夠及不能夠購得的產品。換句話說，在順序上是先說明理由之後，再陳述請求事項。

然而在改善範例中，段落的順序為：⑴希望在當地購買危險物品，所以請對方告知在當地可以購得的危險物品；⑵理由是因

為……。從順序上而言，是先說明要求，請對方做哪些動作，之後再說明理由。

像原文這種順序——先陳述理由後再提請求，這是寫信者很自然的邏輯；但是對讀信者而言，會產生到底要我做什麼的疑問，而不得不將英文信讀完。改善範例則是先表明希望讀信者採取的行動，讓對方讀了第一段之後便能夠放心地將信讀完。為了以此順序書寫，就必須將自己的邏輯順序前後對調。

這種以讀信者為出發本位的順序叫做 audience-oriented sequence，是一般商務書信的大原則。（第Ⅷ章「提案」第 3 節的參考中，會介紹各式各樣的順序。）

3.　書寫的內容要讓閱讀者能夠了解

很多英文書信或傳真的意思難以理解，閱讀者雖然可以根據前後文來推敲，但是書寫的內容應該讓閱讀者能夠一目瞭然。為了達到這個目標，使用的字彙必須是閱讀者也能了解的，而且要有條理，文法上也不能有嚴重的錯誤。有些人主張不必太拘泥於文法，但是絕對要避免會使閱讀者誤解內容的錯誤。為了避免文法上的嚴重錯誤，就得從字典中確認用法。

下面這篇例文有多處的錯誤和語焉不詳的情況，讀者必須考慮前後文，才能猜想書寫者到底想說什麼。

例文 18　委託調查是否需要安全證明書（原文）

Re: Safety Approval

Would you please investigate whether the certificate of safety approval is needed on customs clearance of new models in your country hereafter or not by the following reason.

Now, we are considering to summarize the version of our new models scheduled to be introduced from 199x. In case of this consideration, we have to also consider the correspondence to the safety approval.

Although we have prepared the certificate of the safety approval for Spain and you on customs clearance of our new models, regarding the case of Spain, now we have to prepare only maker's statement mentioning that the proper model is in conformity with regulation of IEC 65. Please refer to the attached copy of this statement for Spain. (We have received information that even this statement may be unnecessary from 199x.)

So, would you please investigate the case in your country regarding this matter and let us know the actual situation for our information.

Your understanding and cooperation would be much appreciated.

【問題點】

　　第一段可能要表達的意思是：「為了要在貴國辦新型產品的通關，請調查今後是否需要安全許可證明，理由如下：」

　　第二段 In case of this consideration，因為前一句是 we are

considering... （我們正在考量），所以應該是「為了這層考量」的意思。

第三段 for Spain and you, 意思可能是 for Spain and for your company （為了西班牙與貴公司）。

最後一行的結尾是我們最喜歡使用的方式，但是毫無意義，只不過是書寫者無意識地加上一筆而已。或許是為了表示禮貌，但卻是單向而且為上對下的講法。

例文 18 的改善範例 （○）

Re: Required Customs Documentation about Safety

① Could you please find out whether a certificate of safety approval is required for customs clearance[1] in Spain? Also, could you find out whether such requirements will be changing in the near future? We are now planning the documentation for the models to be introduced in 199x.

② The reason I make this request now is that some countries have recently changed the customs clearance requirement regarding safety certification, so that just the maker's declaration[2] that the model is in conformity with[3] IEC 65 is now required.

Best regards,

Paul King

---

1) customs clearance　通關　2) declaration　（通關的）申報單　3) in conformity with~　適合~

【段落大綱】

① 陳述委託調查的要點及其原因、目的。
② 詳細説明委託調查的原因及目的。

【內容】

主旨：安全性相關的通關必要文件

請代為求證西班牙通關時是否需要安全許可證明，而且短期內這個規定是否會改變。我們計畫在 199x 年引進的新型產品附上安全許可證明。

之所以提出這個請求，是因為近來有些國家在通關時安全許可證明的必要性已經有所改變，他們目前只要求廠商提出申報單，證明此型產品符合 IEC 65 即可。

【注意事項】

跟原文相同，第一段陳述請求的要點，極力地拜託，所以有時候甚至可以省略第二段。

一般人的演講經常從分論開始，一旦時間不夠無法講完時，就令人不明白他想說些什麼。如果能夠開宗明義，即使在中途打斷，演講還是有整體性。這篇例文就是相同的情形。

4.　資訊的組織

商務書信不可以想到哪寫到哪，需依據資料整理內容、做出結論後，決定要傳遞什麼樣的訊息給讀者，以及要如何達成。此外，還得考慮用什麼樣的邏輯和順序來告知讀者。這裡所謂的邏輯，一般是由眾多要素和許多階段組成的，它不是平面的，而是有層次的。這種有層次的邏輯所組成的資訊，經過有效地組織，

成為易懂的文字，這樣才算是一封書信的大功告成。

　　如果沒有顧慮到提示資訊的順序，或是資訊的組織凌亂，想要表達的事情就無法傳達到讀者心中，而無法達到預期的效果。

例文 19　委託研究新材料的傳真（原文）

Apr. 19, 199x

Mr. H. Young
International Procurement Division
XYZ Company

Re: Mouse Cover Material

We would like XYZ to evaluate materials sample, which Mr.
Owen provided to Mr. Bill Green on Apr. 6.

1. ABC was not selected as new mouse vendor.
   However, current mouse will continue at least by the end
   of 199x. (It will be 150 kpcs on the 2nd half of 199x.)

2. DEF Electric (current material vendor) has increased material cost recently.

   We would like to maintain current cost for current mouse.
   Therefore we would like to propose strongly that XYZ
   re-evaluate this material. And we would like to approach

---

new mouse (Case B) with this material.

Regards

A. Taylor
Marketing Dept.

File: MOUSE4

【內容】
　　<有些地方實在文意不明，推測應該是如下的意思。>
　　希望 XYZ 公司研究一下（本公司）歐文先生於 4 月 6 日交給（貴公司）比爾‧格林先生的材料樣本。
　　1.本公司沒有被選為新滑鼠的承包商，但是目前滑鼠（的生產）至少還維持到 199x 年年底，（其生產量）199x 年度下半年約為15 萬個。
　　2.現在的材料供應商 DEF 公司電器近來漲價，（可是）我們仍希望維持目前滑鼠的價格。所以，我們建議您重新研究此一材料（新材料）（決定採用目前之滑鼠）。另外，B 案的新型滑鼠也準備使用新材料。

【問題點】
　　像上述這個例子要靠推測才能想像內容的原因，在於當事者之間雖有共識，但是省略了能幫助第三者理解的訊息。如果這只是侷限於某些人之中的溝通，難免會產生這種問題；但是這篇例文的段落結構不清，也是令人難懂的原因之一。
　　這份傳真的段落問題點，在於句子都是機械性的換行，不曉

得這是句子的結束或是段落的結束。所謂的段落，是針對同一主題的句子群；所以如果把段落分得過細，原本應該歸納在一個段落的主題就會被分成好幾個段落，邏輯就會顯得鬆散。

　　這份傳真的第一項和第二項中，有些內容應該放在同一段，有些則應該分為另外一段。改善範例就比較容易看懂，也較具說服力。

例文 19 的改善範例（○）

Re: Request to Consider New Mouse Cover Material

① We would be grateful if XYZ would evaluate the material samples provided by Mr. Owen of ABC to Mr. Bill Green on April 6.

② We propose to use the new material for the remainder of your order[1] for the current mouse (150 kpcs. in 2nd half of 199x). The current material vendor has increased prices. Since we want to maintain present prices on the current mouse, we suggest the new material.

③ We are sure from our own tests that this material can be used with no deterioration in quality[2]. In fact, we expect to recommend this new material for the new mouse (Case B).

Sincerely,

1) the remainder of your order　貴公司訂單剩餘的部分（尚未出貨的部分）　2) deterioration in quality　品質惡化

【段落大綱】

① 請求研究先前給的材料樣本。
② 說明希望採用此一材料的原因。
③ 預告該材料品質優良，也準備運用到其它產品上。

【內容】

　　主旨: 請求研究滑鼠外殼材料

　　請求貴公司研究本公司歐文先生交給比爾‧格林先生的材料樣本。

　　我們建議目前尚未出貨的滑鼠（199x 年下半年還有 15 萬個）使用這個新材料。目前材料廠已經漲價，但是本公司仍希望維持現有滑鼠的價格，特別推薦使用這個新材料。

　　本公司實驗的結果，確信使用此一材料不會造成品質惡化。事實上，我們也建議新滑鼠（B 案）使用這個新材料。

5. 並非一味地保持客氣和禮貌就是好的

　　我們是很講究禮貌的，有的人認為這種禮貌是一種美德，但是也不用過度有禮。上司對下屬的命令、顧客對業者的指示，或是買方對賣方的抱怨等，太過有禮貌反而不自然。英文的禮節彙整於下一節。

例文 20　委託辦理簽證手續的傳真（原文）

Sept. 26, 199x

To:　　Mr. D.T. Lin
　　　　Taiwan Technical Service

From:　T. Adams
　　　　ABC Company

Re: Mr. Mason's Letter of Appointment[1]

We will send you the original letter of appointment on September 27th by DHL[2], as the FAX attached herewith.
(2 copies, legalized[3] in Association of East Asian Relations)

Please rush to[4] start Mr. Mason's VISA procedure.
Please confirm.

Regards

1) letter of appointment　決定拜訪日期的信件　2) DHL　國際快捷簡稱　3) legalized　經過正式證明的　4) to rush to ～　急著去做～

【問題點】
　　這封信整體上都給人性急的感覺。如果時常拜託對方幫忙辦簽證手續，那麼實在沒有特別講究禮貌的必要。
　　最後一行 Please confirm.　一般用來表示「告知行動的結

果」，to confirm 的意思是「確定如此」，也就是「確認」。「確認」和英文的 to confirm 多用於下達指令，而指令一定要正確地傳達。

例文 20 的改善範例 (○)

Re: Mr. Mason's Letter of Appointment,
　　in Support of Visa Application[1]

We have sent original letters of appointment (two copies,
legalized by Association of East Asian Relations) by DHL,
to arrive there on September 27. On receipt, please begin
immediately Mr. Mason's visa application procedure. Please
inform as soon as application has been made.

Best regards,

---

1) visa application　申請簽證

【內容】

主旨：梅森先生請求協助申請簽證的預約信

我們以 DHL 寄送預約信正本（二份，經東亞關係協會證明），預計於 9 月 27 日送達。收到後，請即刻開始辦理梅森先生的簽證手續。手續完成後請立即通知。

【注意事項】

第一句就提到文件到達的日期，這樣的寫法非常體貼。第三句的「手續完成後請立即通知」也非常具體，不像 Please confirm. 那樣模糊。

6. 「請～」的句型

寫信的目的，是要讓讀信的人能夠知道某件事情，或是說服讀信的人採取某種行動。日常的商業書信中，說服對方做某事的情況要比告知對方的情況多很多。

說服型的書信最常用的表達方式是「請～」，如「請回答」、「請考量」、「請多多指教」「請多多關照」等。英文中相當於「請～」的有 "We would like to request you to..." 等，並非僅有「Please ＋ 命令式」。

英文中相當於「請求」的客氣說法有幾種呢? 我們來看幾個例子。例文中「原文」是國人所寫的; （△）是經由以英文為母語者修改過文法的錯誤，或是改為英文式的表達方式; 而「修改範例」（○）則是著重禮貌而重新改寫過的。

【背景】

某企業與海外夥伴合作，向客戶提出汽車零件製造廠的提案，客戶卻要求考慮能夠回銷。企業為了得知合作夥伴的看法，於是發了這份傳真。

例文 21 「煩請告知您的看法」（請求考量）（原文）

> Counter purchase (C/P) for automobile parts produced in this plant is requested by the clients in this project.
> *Please inform* your basic concept for this counter purchase to us by end of June, 1991.

第一句寫 Counter purchase (C/P) 表示以後的 counter purchase 都會以 C/P 的縮寫來代替。所以第二句的 counter purchase

應該寫成 C/P，這才是縮寫的正確用法。

第二句對 A（人）通知 B（事），應該是 inform A of B，而非 inform B to A，所以這裡應該寫成 inform us of your basic concept。

例文 21 以英文為母語者修改的範例（△）

> Counter-purchase of automobile parts produced in this plant is requested by project Client. *Please let us know* your basic concept for this counter-purchase by the end of June, 1991.

Please let us know 是祈使句，但是命令的對象不是對方而是自己，所以比 please inform us 要客氣得多。

例文 21 的改善範例（○）

> Client for this project requires counter-purchase of automobile parts produced by this plant, and has urgently requested information to arrive by July 7. *May we ask you to have* a basic outline of such counter-purchase possibilities *sent* to us, to arrive here no later than the end of this month? I ask this so that we may translate and forward it by the deadline given us by this client.
>
> We regret late notice but hope with your help to promote the awarding of this project to our group.

【內容】

　　本計畫的客戶要求此工廠中生產的汽車零件回銷，要我們趕

在 7 月 7 日之前回答。請在本月底前,將可否回銷的基本大綱寄給我們,因為我們要在客戶提出的日期前翻譯好後寄出。

很抱歉,這麼晚才通知您,希望在您的合作之下,這個計畫也會發包給我們的團隊來進行。

【注意事項】

我們來看看改善範例第二句「請告知」的英文有哪些說法:

(1) Please inform us of your basic concept. (原文)

(2) Please let us know your basic concept. (以英文為母語者修改的範例)

(3) Could you inform us of your basic concept?

(4) May we ask you to inform us of your basic concept. (改善範例的變化)

(5) May we ask you to have a basic outline of possibilities sent to us. (改善範例)

(1)與(2)幾乎相同,而(3)、(4)、(5)是越來越禮貌。改善範例中用了 have 這個使役動詞,將要求對方的感覺變得和緩,同時還充分說明了為什麼一定要在月底之前的原因。

改善範例中的 no later than 與 not later than 是不一樣的。「no+比較級+than」如果換成「as+反義的原形+as」來考慮,就更容易懂了: no later than the end of July=as early as the end of July.

【背景】

下面是典型的委託估價單 (request for quotation,縮寫為 RFQ) 的一部分。許多公司都是利用這種 RFQ 委託海外公司報價,但是另有更親切的表示方法。

例文 22 「請求報價」（委託估價）（原文）

> ...This is *requesting* you to submit a quotation for a part of such equipment. *Please furnish us with* your quotation in four copies in English under the following terms and conditions as well as the attached Requisition.

第二段 Please furnish us with 是「請將～交給我們」的意思。同句中 terms and conditions 是「（交易）條件」，terms 和 conditions 意思一樣。在法律條文中，向來有把法語系的單字和英語系的單字重複使用的習慣。法律用語重複的例子還有 by and between, each and all, final and conclusive, null and void, made and entered into 等。

requisition=official demand，是相當於「採購申請書」的單字。

例文 22 以英文為母語者修改的範例（△）

> ...This is *to request you* to submit a quotation for a part of such equipment. *Please furnish us with* four copies of your quotation in English, under the terms and conditions given below and in the attached Requisition.

第二句 in English 之後加上逗號，表明以下片語並非用來修飾 English，而是在修飾 furnish。

最後的 the attached Requisition 前面加上 in，也是用來明確表示 given in the attached Requisition 的意思。

例文 22 的改善範例（○）

> ...We are happy to *invite you* to make a quotation for a part of such equipment. If you are able to make such a quotation, *may we please have* four copies, in English, following terms and conditions given below and in the attached Requisition.

「請求估價」的動詞並非只有 to request 或 to ask，就像這篇改善範例開頭部分一樣，可以使用更客氣的 to invite。第二句的「請提出」，使用 may we (please) have～比 please furnish us with～要有禮貌。

【內容】

請針對（上述）機器的部分做報價。如果您能提供報價，請依照下述條件及附件的採購申請書所示條件，以英文提出四份。

【背景】

接下來的例子是某企業想要變更指定機器的配置，並請客戶許可變更。

例文 23 「請認可」（請求認可變更）（原文）

> ...We would like to propose the adoption of a configuration different from your original idea.
> *We would very much appreciate* your review over the following consideration and *like you* to give us your approval by the first week of February for our present design to make it our mutual design philosophy.

第二句 We would very much appreciate 和 we like you to～都是「請求」的用法。最後的 to make it our mutual design philosophy，意思是「將它作為貴我雙方的設計理念」。

例文 23 以英文為母語者修改的範例（△）

...We would like to propose the adoption of a configuration different from your original idea.

We would very much appreciate your review of our consideration, on following pages, and *would like you to* give us your approval for our design by the first week of February, so that this design can become our mutual design philosophy.

修改原文中的文法錯誤或語焉不詳之處。

例文 23 的改善範例（○）

...In this particular application, a different configuration than that given in the specification appears to offer advantages, and we propose the adoption of this configuration for this project.

We *would very much appreciate* your review of our investigations and proposal given in the following pages. *We are hoping that* you will be able to give us approval by some time in early February, so that we can continue work on this project.

第一段 configuration 是「配置、結構」的意思。第二段in early February是「2月上旬」，另外還可以用 early in February， at the beginning of February, in the first part of February 等方法表示相同的意義。

【內容】

此時，不同於規格書所示的結構似乎比較有利，因此建議本案採用此結構。

請研究下頁所示本公司的調查結果和提案。為了能讓我們繼續進行此案，希望能在 2 月上旬前獲得您的許可。

【注意事項】

改善範例中沒有談到 mutual design philosophy，只要求一次許可，而且期限也寫得比較委婉。《美國銷售手冊》中有句格言 "You can't lose what you don't have." 如果請求許可是一開始就成功機會渺茫的話，更應該努力達到效果。

【背景】

某建廠業者由於無法從工廠興建地的埃及客戶獲得土質資料，所以交不出高塔的基礎圖。於是決定推算土質數據，再依據推算來繪製基礎圖。為了要讓客戶認可這些推算值，於是發了下面的傳真。

例文 24　「請認可」（請求認可）（原文）

We regret to inform you that we cannot submit the tower foundation drawings within November.
This is because we have not received the actual soil date[1] for each station to date.
However, we appreciate your circumstances and intention to complete the project as early as possible.

Therefore, we propose that we design the foundations based on the following assumptive conditions.

（推算值）

We hope to receive your official approval for the above by return mail.

Your approval should clearly indicate that you are fully responsible for the above assumptive soil conditions.

Your kind understanding and quick action *would be highly appreciated*.

1) soil data　土質資料

【內容】

　　我們很遺憾無法在 11 月中提出高塔的基礎圖。

　　因為目前我們沒有收到各站用的實際土質資料。

　　但是我們了解您的狀況以及希望儘早完成本案的心意。

　　所以，我們建議依據下列推算條件來進行基礎設計。

　　　　　　　　（推算值）

　　希望獲得您正式的認可。

　　您的認可中應明確說明上述推算土壤條件，您將負起全責。

　　感謝您的體諒及迅速的處理。

【問題點】

　　要客戶對推算的數據負責，實在有點強人所難，但如果是合約上有明定，那就可以這麼要求（參照改善範例）。

　　原文的問題點在於每換一個句子就換行。換行不應在換句的時候，而是在變換段落的時候。此外，在更換段落時，要不就隔一行，要不就在開頭時要留空白（約 5 個 stroke）。每換一句就

換行，讀者會以為是換了一個新的段落。

例文 24　以英文為母語者修改的範例（△）

> We regret that, because we have not yet been able to obtain soil data from some of the stations, we will not be able to submit tower foundation drawings before the end of November. However, we appreciate your circumstances and your intention to complete the project as soon as possible. Therefore, we propose that we design the foundations according to the following assumptions.
>
> （推算值）
>
> *We hope to receive* your official approval for these assumptions by return mail. Your approval should clearly indicate that you are fully responsible for these assumptions regarding soil conditions.
>
> Your kind understanding and quick action *will be highly appreciated*.

　　倒數第二句的 Your approval should clearly indicate～的 should 雖然沒有 shall 或 must 那麼帶強迫性，但力道還是很強。也就是說，不論原文或修改範例，兩者都不是在請求，而是在強求。

例文 24 的改善範例（○）

> We are sorry to report that, because of the continuing un-availability of soil data from the sites of the stations, we will not be able to submit tower foundation drawings before the end of this month, as we had planned.
>
> We realize, however, that you are as anxious as we are to complete the project as quickly as possible. We would like to suggest for your consideration, therefore, if you are not able to supply the actual soil data, the following assumptions regarding soil conditions. If you approve of them, we will design the foundations based on them.
>
> （推算值）
>
> *We hope that you are able to give* these proposed soil as-sumptions your consideration very soon. *We will very much appreciate it if you can* give us your response within this week, because the delivery of materials has already been arranged and many of the materials are already in transit. (Naturally, as stipulated in the Contrace, Paragraph xxxx, these soil con-dition assumptions will be considered as fully and completely yours, with all responsibility for them being yours also.)
>
> *We are looking forward to* your early reply, and the resumption as soon as possible of work on the project.

　　（推算值）前面的 approve of～是「同意～」的意思，與 ap-prove～（認可～）不同。（推算值）後面第二句相當於「請求」

的表達方式，用的是 We will very much appreciate it if you can～，這樣比較客氣。而關於推算數據的責任問題，則委婉地寫在括弧內，而且引用合約。最後一句的 resumption 是「重新開始」的意思。

【內容】

　　由於遲遲未能收到各站的現場土質資料，所以很遺憾地，無法如先前計畫於本月底提出高塔基礎圖。

　　但是我們了解您和本公司一樣，殷切期盼本案能夠儘早完成，所以如果您無法提供實際的土質資料，建議您考慮下列關於土質條件的推算。

（推算值）

　　我們希望您能儘早研究我們提議的土質推算值。我們已經安排好材料的交貨，許多材料已經在運輸途中，所以希望能在本週內獲得答案（如合約書中第 XXXX 項的規定，土質條件的推算值當然完全歸您所有，而您也必須負起全責。）

　　期待您盡快回覆，以及本案作業能盡快重新開始。

（參考）有禮貌的「請求」

　　在此將本節討論的「請求」的英文表達方式做一個整理。下面的例子，由上而下，是越來越有禮貌。但這個順序也不是絕對正確，要看對方和前後文而定。而且也不是一味地客氣就是好的。用法不對，太過客氣反而失禮。

1. 命令式：不加 please 的命令句語氣非常強烈。（但是手冊或警報標語等命令句不加 please。）

　　例：Send it to me by return mail.

　　　　（回寄給我。）

2. I/We want/expect you to～　強迫對方接受自己的希望。用於上司對部屬下命令時。

　　例：I want you to complete the work by tomorrow.
　　　　（我要你在明天之前完成這個工作。）

　　例：We expect you to ship the order by the end of November.
　　　　（希望你在 11 月底前將訂單出貨。）

3. Please＋命令形：即使有 please，命令的口氣不變。

　　例：Please send it to me by the end of this month.
　　　　（本月底前寄給我。）

4. I/We'd like you to～　雖然客氣，但是強迫對方接受自己的希望。

　　例：I'd like you to give us your approval be the end of this month.
　　　　（本月底前請認可。）

5. Let me (us)～　雖然是命令形，但在要求許可，所以要比4.客氣很多。

　　例：Please let me have your order by July 31.
　　　　（請在 7 月 31 日前下訂單。）

　　例：Let me get straight to the point.
　　　　（恕我直言要點。）

6. I/We would request you to～
　　I/We would request that you～

　　或許很多人都以為 request 相等於「請求」，所以經常使用，但這是比較打官腔式的強硬用法，最好要小心。不用 request 而用 ask，例如 we ask you to～或是 we would ask that you～，基本上也沒什麼太大差異。

　　例：I would ask you to reconsider this matter.

（請重新考慮這件事。）

例：We request you to pay the amount immediately.

（請立即支付款項。）

7. It would be appreciated if～　雖然有禮貌，但因為是被動式，所以有些打官腔的味道。

例：It would be appreciated if you would send it by return mail.

（如果能寄回，萬分感激。）

8. I/We would appreciate～　是有禮貌的表現，常用於書信等中。但是下面的 I would appreciate it if～過於客套，給人反諷的感覺。

例：We would appreciate your review of our consideration.

（如蒙考慮我方之想法，萬分感激。）

例：I would appreciate it if you would send it by return.

（如蒙寄回，幸甚。）

9. Would you (please)～?
　Could you (please)～?

口語且自然的表達方式，而 could 又比would 委婉。

例：Would you mail us two copies of your invoice, please?

（能否寄來二分發票?）

例：Could you please send it to me as soon as possible?

（能否儘早寄回?）

此外，加強客氣的說法還有下列幾種：

1. 躊躇的表達方式：I wonder, you might be able, could it be possible 等。

例：I wonder whether I might ask your help.

（能否請您幫忙？）

2. 使役動詞： 藉著認定對方地位高來表達敬意。

例： I wonder if you might have one of your staff send it to me as soon as possible.

　　（能否請您的屬下早日寄來？）

即使知道對方沒有屬下，還是可以使用這種說法。沒有人會因為自己的地位被提高而生氣。

3. 說明理由： 清楚說明理由也是肯定對方的做法，能增加說服力。但這個理由可不能光是出於書寫者的立場，還要與閱讀者有切身關係。

例： I will be unable to complete the drawings until you have send me the photos, so could I ask you to have them within the next three days?

　　（您如果不寄照片過來，我無法完成設計圖，能否在 3 天之內寄來？）

催
促

1. 尊重對等主義

一國外交部長的正式會談對象當然是另外一個國家的外交部長，而要平等接待某國元首的訪問，自然是要被訪問國家的元首親迎。這種正式的會談或款待，必須由地位相等的人員來進行的原則叫做對等主義 (reciprocity)。對等主義不僅在外交上，一般民間也是遵守這個原則，總裁對總裁、經理對經理，這都是正式來往時所須遵循的原則。但是，因為公司的規模、買賣雙方的關係、對等職位的不同，關係自然有些差異。舉例來說，客戶 A 公司的副總裁可以由賣方 B 公司的總經理接待。

下面的例子，是買方（機械工廠）的經理向賣方（零件廠）的總裁抱怨延遲交貨的電報。

【背景】

ABC 公司向美國 XYZ 公司下單，訂購客戶 DEF 公司用機械零件。ABC 公司的李經理最擔心對方延遲交貨，於是在下單之際，要求 XYZ 公司的史都華總裁簽署一份保證不延遲交貨的信函給 ABC 公司的懷特副總裁。李經理更透過 XYZ 的香港分公司，定期確認生產情況。

某日，XYZ 香港分公司向李經理報告進度延宕五到六週。詫異之餘，李經理直接打了一份電報給史都華總裁。

例文 25　抗議延遲交貨（原文）

```
ATTN: MR. M. STEWART, PRESIDENT OF XYZ USA
COPY: MR. W. SMITH, VICE PRESIDENT OF XYZ
      HONG KONG
```

WE ARE VERY SURPRISED TO BE INFORMED BY XYZ HONG KONG THAT THE MANUFACTURING SCHEDULE OF OUR PARTS IN YOUR SHOP IS NOW 5 TO 6 WEEKS BEHIND THE CONTRACTED SCHEDULE[1].

WE MUST SAY THAT 5 TO 6 WEEKS DELAY AT THIS EARLY STAGE OF MANUFACTURING IS A SERIOUS PROBLEM FOR OUR COMMITMENT TO OUR CLIENT[2].

PLEASE REFER TO YOUR STATEMENT IN YOUR LETTER OF APRIL 5, 19xx, TO TOM WHITE, ABC'S VICE PRESIDENT, AND BE DULY REQUESTED TO TAKE BY YOURSELF ALL MEASURES NECESSARY FOR IMPROVING THE PROGRESS TO ENSURE THE CONTRACTED DELIVERY MONTH OF NOVEMBER 19xx.

WE THINK THE ALLOCATION OF YOUR MACHINERY AND WORKERS TO THIS PROJECT IS MOST IMPORTANT.

BEST REGARDS,

S. LEE, PROJECT MANAGER

---

1) be behind the schedule　比計畫落後　　2) commitment to our client 對客戶的約定

【內容】

　　接獲貴公司香港分公司通知，本公司委託貴公司的零件生產進度比合約中預定的落後 5 到 6 週，實在令人驚訝。

　　機械加工的初期階段就產生 5 到 6 週的延宕，在與客戶的約定上，不啻是個嚴重的問題。

　　請參照您 4 月 5 日寄給本公司湯姆‧懷特副總裁的信函。為了確保約定今年 11 月的交貨期，請您親自督導，採取各項必要措施，以改善進度。

　　我們認為多分配您的機械與員工到此一專案中是最重要的。

【問題點】

　　首先是用法上的問題，第三段 be duly requested to... 的用法過於官僚。request 這個動詞本身就帶有官僚氣息，再加上 duly 更是官僚之至。此外，第一段 We are very surprised...則不僅表示訝異，還帶有不悅的語氣。

　　本節一開始就提到了對等主義，所以 4 月 5 日的信件既然是 XYZ 公司的史都華總裁寄給 ABC 公司的懷特副總裁，此時與賣方 XYZ 公司總裁地位等同的對方，應該是買方 ABC 公司的懷特副總裁。在這樣的前提之下，李經理直接打電報給史都華總裁，是不諳規定的做法。另外，「請總裁您親自督導，採取各項必要措施」（第三段）也是很不禮貌的說法，請和改善範例做個比較。

　　後來，XYZ 公司對於李經理電報的回應，是由香港分公司的史密斯副總裁回覆的，似乎是在告訴李經理，「你的對象應該是香港分公司的副總裁才是！」

例文 26　回覆交貨期（△）

Dear Mr. Lee:

In accordance with your request of 8/22/9x in our meeting, we confirm that the subject orders[1] will be ready for shipment, ex-works[2], Ann Arbor, Michigan, on the following date:

Main order	11/12/9x
Additional order	11/19/9x

We trust this meets your satisfaction.

Sincerely,

William Smith
Vice President
XYZ Hong Kong

---

1) subject order　標題所示（但這裡省略了標題）的訂單　　2) ex-works　工廠出貨

【內容】

依據您 8 月 22 日會議中所提出的要求，我們確認上述訂單將在下述日期於密西根州安雅柏工廠完成出貨準備。

主要訂單　　　11 月 12 日
追加訂單　　　11 月 19 日
相信您一定能滿意這個交貨期。

【注意事項】

XYZ 公司的回信中廢話也不多說，只說結論。而交貨期的問題，雖然現在才 8 月，也不管到 11 月前會發生什麼事，他們就保證 the subject orders will be ready for shipment...on the following date.

這個助動詞 will 非常重要，如果不用 will 而用 may 或can 的話，11 月的交貨期又要令人擔心了。

11 月 12 日或 11 月19 日的日期，是以週為考量單位而得出的結果。而我們習慣以月工作為考量的單位，有時候會和歐美人的認知有落差。

李經理跳過史密斯副總裁的層級，直接打電報給史都華總裁，在回函中也沒有再提及此事，只把它當作是李經理與史密斯副總裁之間的事務來處理，這彷彿就是在確認李經理商務上的對等者應該是史密斯副總裁，是一篇簡潔且典型的商業信函。

但是，像例文中 8/22/9x 與 11/12/9x 等日期的表示方法請勿採用。

1994 年 11 月 12 日，美國一般的寫法是 November 12, 1994，數字則是 11/12/1994；英國寫成 12 November 1994 或 12th November 1994，數字則是 12/11/1994。例文中只有數字的日期表示方法，非常容易造成誤會。

接下來介紹原文的改善範例。

例文 25 的改善範例（○）

WE WERE WORRIED[1] TO HEAR FROM XYZ HONG KONG THAT MANUFACTURING OF THE PROJECT PARTS IN YOUR SHOP IS NOW 5 TO 6 WEEKS BEHIND THE SCHEDULE[2] IN OUR AGREEMENT. SUCH A

SUBSTANTIAL DELAY[3] AT THIS EARLY STAGE MAY CAUSE A SERIOUS STRAIN[4] BETWEEN ABC AND OUR CLIENT.　I RESPECTFULLY REMIND YOU OF[5] COMMITMENTS[6] MADE BY XYZ TO TOM WHITE, ABC'S VICE PRESIDENT, AND URGENTLY REQUEST THAT YOU HAVE YOUR STAFF TAKE ALL MEASURES[7] NECESSARY TO ADHERE TO[8] AGREED-UPON DELIVERY[9]. WE BELIEVE THE KEY IN ACHIEVING THIS TO BE YOUR ALLOCATION[10] OF EQUIPMENT AND WORKPOWER[11] TO THIS PROJECT.

1) be worried　擔心　　2) is behind the schedule　比預定進度落後
3) substantial delay　重大延誤　　4) strain　原本是緊張或疲勞的意思，在此用作「不信任」(state of difficulty, distrust, opposition between people or groups, *LDCE*)　　5) to remind A of B　讓A想起 B
6) commitment　承諾、約定　　7) to have your staff take measures 要您的下屬採取措施　　8) to adhere to～　固守～　　9) agreed-upon delivery　同意的交貨期　　10) allocation　分配　　11) workpower　勞動力（與 manpower 意思相同，但是 manpower 帶有 man 這個字，所以有人認為這是性別歧視用語 (sexist word)）

【內容】
　　由香港分公司知悉貴廠機械零件的生產進度比合約中落後 5 至 6 週，實在令人擔憂。初期階段就發生這麼嚴重的延宕，可能會造成本公司與客戶之間的大問題。請回想 XYZ 公司對本公司懷特副總裁所做的承諾，迫切地請您指示下屬採取必要措施，以遵守合約期限。我們相信目標要達成的最重要關鍵，就在於為本

專案多分配設備與人力。

【注意事項】

我們還是依照原本的做法，由經理打電報給總裁，但用的是非常有禮貌的語氣。舉例而言，原文是 (please) be duly requested to take by yourself all measures（請總裁親自採取各項措施），而改善範例則是I urgently request that you have your staff take all measures（迫切地請您指示下屬採取措施），是截然不同的思考方向。「請總裁親自採取各項措施」，說不定會惹怒總裁；但是「請您指示下屬」，可能對對方的自尊心產生正面效果。這種表達方式，對於那些不知道有沒有下屬的中、下層級主管，或是沒有下屬的專業人員，也都可以使用。利用希望獲得別人肯定的心理，是工作成功的祕訣之一。

2. 說明理由

ABC 工程顧問公司將自行製作的 UVW 公司（中近東）工廠改建的可行性報告書寄給 XYZ 協會（印度），接受進一步的確認。XYZ 協會應將 ABC 公司報告書令人滿意的認證書以及該報告書一併寄給 UVW 公司，如此 ABC 公司才能獲得UVW公司支付可行性調查報告的酬勞。

到目前為止，ABC 公司接到 XYZ 協會寄給 UVW 公司的附帶信函影本，確定 XYZ 協會已經將 ABC 公司的報告書寄給 UVW 公司，但是卻不確定 XYZ 協會有沒有附上認證書，於是 ABC 顧問公司的楊經理發了一份傳真給 XYZ 協會。

例文 27　催促簽發文件（原文）

13th April 19xx

Attn: Mr. P. K. Singh, General Manager
Re: Feasibility Report to UVW

Dear Sirs,

We acknowledge hereby the receipt of a copy of the Feasibility Report that you submitted to UVW.

Please refer to our telex No. FRX–011 dt. 11th Apr. 19xx to Mr R. Krishnan (Chief Eng'r). We would like to ask you again if you have issued a certification to UVW. If so, please send us a copy of it. If you have not issued it yet, please make prompt arrangement of issuance of the certificate.

Thanking you,

Very truly yours,

K. Young
Project Manager

【內容】

　　我們在此感謝收到您寄給 UVW 公司的可行性報告書影本。
請參照本公司於 19xx 年 4 月 11 日發給克里詩南先生（總

工程師）的電報，號碼 FRX-011。再次請教是否已經發出認證書給 UVW 公司了呢？如果是，請寄一份影本給我們。如果尚未發出，請立即發出認證書。

【問題點】

從這份傳真可以看出「為什麼不寄認證書過來？」的怒氣，不僅內容相當公式化，也給人強悍的感覺。

本文第一句 We acknowledge hereby the receipt of... 太過僵化，用 We have received... 就可以了。

第二句...dt. 11th Apr. 19xx 是不能簡寫的，要寫成...dated 11th April 1993；而且也不是 dated，而是 of 比較好。此外，因為不是電報，所以不能縮寫為 Chief Eng'r，要寫成完整的 Chief Engineer。

最後一句的 Thanking you 也是過於陳腐的用法。

例文 27 的改善範例（○）

Dear Mr. Singh:

Re: Issuance of Certificate of Completion of Feasibility Study[1] to UVW

① Thank you for the copy of the Feasibility Report, which we received on April 16. We also urgently require a copy of the Certificate of Completion of Feasibility Study, as issued by you to UVW. If that has not been issued yet, we request that it be done[2] by April 30, and a copy sent to arrive at our office by May 5.

> ② As we said in our telex of April 11, preparation of the foundation drawings cannot begin until this document has been received here. We appreciate your earliest attention to this urgent requirement.
>
> Yours very truly,

1) feasibility study　計畫是否確實可行, 也就是要決定應否開始進行計畫的「事前調查」, 該報告書稱作 feasibility report　2) we request that it be done　也可以說成 we request that it should be done, 但很少人用should, 而往往採用例文的形式。

【段落大綱】

---
① 感謝對方寄出可行性報告書的影本, 提示發出認證書的日期。
② 說明立刻需要認證書的理由。

【內容】

　　我們在 4 月 16 日接到可行性報告書的影本, 謝謝您。我們現在急需您發給 UVW 公司可行性調查的認證書。如果尚未簽發, 請於 4 月 30 日以前簽發, 並將影本於 5 月 5 日以前寄給我們。

　　如同 4 月 11 日電報中提到的, 這份文件如果沒有寄達, 就無法著手繪製基礎圖。請以最速件處理。

【注意事項】

在催促簽發可行性調查認證書時，必須說明為什麼需要這份文件，並標明期限。看到這個理由，XYZ 協會也會特別注意才對。

3. 自己人也得注意禮節

跟外國公司做生意，經常為了從對方那邊取得必要文件而大費周章，有時連估價單都無法順利取得。

下面介紹的例子是 ABC 公司機器設計部的梅森主任為了要估算客戶 UVW 公司的提案預算，想從德國廠商 XYZ 公司取得估價資料。可是約定期限到了，卻還不見估價單寄來。絞盡腦汁之後，拍了封電報給該公司歐洲據點的馬丁總裁。

例文 28　催促估價單的電報（原文）

URGENT

ATTN: MR. MARTIN, PRESIDENT OF ABC EUROPE

RE: UVW POWER PLANT–XYZ PUMP

ANY INFORMATION HAS NOT REACHED OUR OF-
FICE FROM XYZ YET. WE WASTE FOUR DAYS AT
PRESENT, BECAUSE IN YOUR LAST TELEX, THEY
PROMISED TO GIVE QUOTATION IN WEEK 50 (DEC
7–11).

WE ARE STRONGLY LOOKING FORWARD TO RECEIV-
ING THE QUOTATION ON DEC 20.

BEST REGARDS

T. MASON
MECHANICAL DESIGN DEPT.

【內容】

　　急件

　　受文: 歐洲 ABC 馬丁總裁鈞鑒

　　主旨: UVW 發電廠用 XYZ 幫浦

　　XYZ 公司至今尚未給我們任何消息。依您最近的電報中表
示，XYZ 公司保證在第 50 週（12 月 7 日—11 日）提出估價
單，但我們已經浪費 4 天了。我們非常希望在 12 月20 日能收到
估價單。

【問題點】

　　上面的英文有好幾處文法的錯誤。此外，可能梅森主任並
不諳於文章的寫法，所以寫出來的英文非常不禮貌。對馬丁總裁
而言，他們也是拼命地想要拿到估價單，卻被說成「浪費了 4
天」，心中想必很嘔。

　　習慣做國內生意的公司，往往會認為只要請對方估價，無論
是哪一家廠商一定都會全力配合，提供估價單。但是和海外做生
意就未必如此了，估價工作是很耗費人力的事。尤其此次發電廠
的案子，ABC 公司也僅止於提案的階段；換句話說，即使 XYZ
公司提出估價單，也不見得就一定能夠接到訂單。等到 ABC 公

司接到案子之後，才是幫浦開始比價的時候，所以很多業者對於無法立即接單的估價請求，往往都不願意浪費人力。也有可能是 ABC 公司或其它本國公司到頭來都還是下訂單給本國廠商，而讓 XYZ 公司或其它海外廠商吃過苦頭，XYZ 公司於是不想為 ABC 公司在提案時提供報價了。也或許是 XYZ 公司已經跟別家公司保證全面提供協助，所以拒絕提供估價單給 ABC 公司。

　　從以上的例子可以知道，以為要求別人估價就一定能夠獲得報價單的想法，其實是錯誤的。馬丁總裁的苦楚，梅森主任可不知情。

例文 28 的改善範例（○）

```
TO   : MR A BECKER, MANAGER
FROM: T MASON, MECHANICAL DESIGN DEPT
RE   : UVW POWER PLANT:  XYZ'S QUOTATION
       FOR PUMPS


WE WOULD LIKE TO REMIND YOU OF[1] XYZ'S
PROMISE TO SUBMIT QUOTATION IN WEEK 50[2],
OR DEC 7–11 (YOUR TELEX OF DEC 1).   COULD
YOU EXPEDITE[3] THE MATTER SO THAT XYZ WILL
QUOTE AT LEAST BUDGETARY PRICE[4] BY TELEX
OR FAX BEFORE X-MAS? ABC MUST COMPLETE OUR
FINAL PROPOSAL BY DEC 26.


REGARDS
```

1) to remind you of～　提醒您～　2) week 50　從年頭開始數第 50 週（歐洲常用的習慣）　3) to expedite　督促　4) budgetary price

預算用的估算價格

【內容】

提醒您 XYZ 公司應允在第 50 週（12 月 7 日－11 日）提供
估價單的保證（您 12 月 1 日的電報）。可否請您督促XYZ 公
司，至少在聖誕節前以電報或傳真的方式提出估算價格。因為本
公司必須在 12 月 26 日前提出定案。

【注意事項】

這封電報的受文者不是馬丁總裁，而是貝克經理。大家或許
會認為同一國家的人，比較容易溝通，但是如果不尊重當地的人
員，企業的國際化就無法向前推展。

4.　催促也得公事公辦

公司對外的收款或文件的延誤，往往會影響到公司內部後
續的作業，所以在催促的傳真中，一不小心就容易摻雜自己的情
緒。但是自己的情緒或焦躁一旦讓對方也感受到，結果一定是負
面的，所以催促時儘量公事公辦較好。

下面的例子是 ABC 機械廠給美國子公司的傳真。XYZ 是
美國的客戶。

例文 29　催促付款的傳真（原文）

XYZ issued P.O.[1] on 1/17/9x and we issued debit note[2] on
3/6/9x according to ABC USA P.O. 0001A (fax 2–15–110). So
we think that ABC USA has already received the amount and

ABC Japan should have been remitted[3] in May 199x based on the ABC Japan debit note issue date 3/6/9x.

It is strange that ABC Japan has not received that amount until July. Anyway we expect the remittance by August.

By the way, please let us know when you invoiced[4] to XYZ.

1) P.O. 訂單（=purchase order）　2) debit note　請款單（參照改善範例的註解）　3) to remit　匯款　4) to invoice　寄出請款單，但 to invoice 也可用於「請款」之意。

【內容】
　　XYZ 公司於 199x 年 1 月 17 日簽發的訂單，本公司依據美國 ABC 公司訂單號碼 0001A（傳真 2–15–110），於 199x 年 3 月 6 日簽發請款單。因此我們認為美國 ABC 公司已經收到該款項。依據 199x 年 3 月 6 日日本 ABC 公司簽發的請款單，ABC 公司應該已經在 5 月就收到匯款。
　　但奇怪的是已經 7 月了，日本ABC 公司並未收到該款項。無論如何，希望在 8 月以前匯款過來。
　　順道一提，請告知我們您是否已經向 XYZ 公司請款。

【問題點】
　　這份傳真本身的文體就不夠禮貌，但因為是母公司對子公司的通信，所以可以不必太計較。然而從接到傳真的子公司立場來看，彷彿自己遭到質疑，心理一定不好受。文中某些地方畫上底線以引起注意，這對讀者也是很失禮的做法。

例文 29 的改善範例（○）

> ...We expect to receive this remittance within the month of
> August. We are puzzled that[1] we have not been able to
> receive it yet. XYZ's P.O. was issued on January 17, 199x,
> and we issued a debit note[2] against the ABC USA P.O.
> 0001A. We have expected remittance based on that debit
> note since May.
>
> Could you please also let us know when you invoiced[3] XYZ
> and when payment was received by you?

1) be puzzled that　因～而困惑　　2) debit note　簿記中的「借方票」，但「請款單」也是這個字。debit note 的反義字為 credit note「貸方票」，當賣方接到退貨時簽發用來扣除請款金額的單據 (*LDBE*)

3) to invoice　名詞 an invoice 是「送貨單」的意思，出貨者在寄送商品時一定要開列的單據。invoice 也可以當作「請款單」來解釋。像是在改善範例中當動詞用，在 invoice 後面加上請款對象，就等於是「向～請款」的意思。

【內容】

　　我們期待 8 月中能收到匯款。到目前為止一直無法收到匯款，至感困惑。 XYZ 公司的訂單是在 199x 年 1 月 17 日簽發的，我們基於美國 ABC 公司的訂單 0001A 開出請款單，從 5 月起就在等候那筆請款單的匯款。

　　能否告知我們您是什麼時候向 XYZ 公司請款？又什麼時候收到付款的？

附件

附件不可過於公文化

附件（letter of transmittal 或 covering letter）是附在寄送文件上的書信。有的附件只有簡簡單單的五、六行，也有比較長的。若要分類的話大致分可為以下三種：

第一種是徹底發揮附件原本的功能，也就是短篇的附件，上頭簡單扼要地說明寄送的文件名稱、寄送的目的、委託對方的處理方式等。第二種是附在文件的最前面，使附件的位置產生最大的效用，比方想要推銷產品、服務，或是塑造公司的形象等，利用這塊地方來加深讀者的印象。第三種是與寄送的文件合而為一，發揮特定的功能。舉例來說，專案的有效期限和折扣原則等，不寫在正式的文件裡，而寫在附件上。

下面的例子，除了是附件之外，對於寄給對方的諸多文件，也在附件中說明處理方法及委託事項。（參照第 V 章第 2 節的例文 27）

【背景】

ABC 工程顧問公司接受 UVW 公司（中東）委託，進行工廠改建可行性的調查。所謂的可行性調查，是在開始推動計畫之前，先評估技術層面、經濟層面是否確切可行。完成後的調查報告書 (feasibility report) 寄送給 XYZ 協會（印度），XYZ 協會將調查業務結束的證明書以及可行性調查報告書一併寄給 UVW 公司，如此 ABC 公司才能獲得約定的報酬。

例文 30　調查報告書的附件（原文）

Pursuant to[1] the "Co-operation[2] Agreement between XYZ Association and ABC Consultants for Preparing a Feasibility Report[3] towards UVW Revamp[4] Project", dated 5th December 19xx, we are pleased to submit herewith "Feasibility Report for UVW Revamp Project" dated 4th March 19xx in five (5)copies.

You are cordially[5] requested to review the report and issue a certificate to UVW Corporation, a copy to us, to the effect that[6] ABC has completed the scope of the study stipulated[7] in the Agreement.

We hope our report will be useful for the implementation[8] of UVW Revamp Project.

1) pursuant to　依據～　 2) co-operation　合作　 3) feasibility report　（參照改善範例註解）　 4) revamp　改建（本來是動詞，這裡當作名詞形的 revamping 來用）　 5) cordially　由衷的　 6) to the effect that　（參照改善範例註解）　 7) to stipulate　明定（合約的條件）　 8) implementation　實施

【內容】

　　根據 19xx 年 12 月 5 日「XYZ 協會與ABC 工程顧問公司間就 UVW 公司改建計畫可行性報告書撰寫的合作協議」，於 19xx 年 3 月 4 日提出「UVW 改建計畫的可行性報告書」影本五份。

請檢閱此報告書後，開立一份證明書予 UVW 公司，一份影本予本公司，說明 ABC 公司已經完成協議書中所明定的調查範圍。

希望此報告書將有助於 UVW 改建計畫的實施。

【問題點】

這封附件就像古文一般，艱澀刻板。

第一段開頭的 Pursuant to 就很八股，若要換成別的說法，可以用In accordance with，或是像改善範例中使用 As agreed upon。

這裡用了兩次的 dated 也很缺乏現代感。「您 5 月 5 日的信」可以寫成(1) your letter dated May 5, (2) your letter of May 5, 或是(3) your May 5 letter。(1)比較舊式，(2)和(3)比較現代化。

同樣地，herewith 這個字，現在連法律條文裡也都不太使用了，意思就像改善範例中 with this letter（連同本信）。不過大部分的情況下，不說 herewith 或with this letter 也不會造成誤會。與 herewith 相近的字還有 hereto（附於本信之後），hereby（根據本信），hereof（本信之），herein（本信中）等，這些都是很八股的單字，一般的書信中最好避免使用。

第二段 You are cordially requested (=You are kindly requested)， cordially 和 kindly 都是表示客氣的副詞，但官腔味十足。

本例文日期的表示方法 5th December, 4th March 是英式的，改善範例用的則是美式的 December 5, March 4。

例文 30 的改善範例（○）

---

① As agreed upon[1] in the contract reached between your-selves and ABC Consultants, on December 5, 19xx, "Co-operation Agreement for Preparing a Feasibility Report[2] towards UVW Revamp Project", we are pleased to submit with this letter our "Feasibility Report for UVW Revamp Project", dated March 4, 19xx. As required in our agree-ment, five copies are enclosed.

② After reviewing the report, may we ask that you issue to UVW Corporation, as soon as possible, as also agreed upon, a Certificate of Performance to the effect that[3] ABC has completed the scope of the study stipulated in the agreement?

③ Your cooperation in the timely conclusion of this feasibility study has been invaluable[4] up to this point, and we look forward to working with you to bring the revamping project to a successful close.

---

1) to agree upon (=on) 同意～，例如 at a price agreed upon 以談妥的價格   2) feasibility report 可行性調查報告書   3) to the effect that 以～為主旨的   4) invaluable 無比實貴的

【段落大綱】

① 說明寄送報告書的主旨以及合約上的根據。
② 請求開立證明書。
③ 感謝以往的協助，並以期待計畫成功做結。

【內容】

　　依據貴協會與本 ABC 工程顧問公司之間所達成的共識，在 19xx 年12 月 5 日訂定「XYZ 協會與 ABC 顧問公司間就 UVW 公司改建計畫可行性調查報告書撰寫的合作協議」，於 19xx 年 3 月 4 日提出「UVW 改建計畫可行性調查報告書」。

　　請於檢閱此報告書之後，依雙方之前的協議，儘早向 UVW 公司提出ABC 公司已經完成協議書中所載調查範圍業務的證明書。

　　本報告書能夠如期完成，多因承蒙貴協會的盛情協助。希望能與貴協會繼續合作，共同促成本改建計畫成功結束。

【注意事項】

　　與八股的原文相比，改善範例較為柔和，並以 may we ask that... 取代原文的 you are cordially requested。

第 VII 章

致意

1. 問候的語氣應考慮對方

　　無論是演講或寫文章，在下筆或開口之前都必須先考慮二件事：一是目的；一是對象（聽眾或讀者）。

　　目的大致可以分為二種：第一種是將消息傳遞給聽眾或讀者；另一種是說服讀者或聽眾採取某種行動。消息的傳遞要讓對方正確地理解；而要說服對方行動，就需要條理分明。

　　考慮聽眾或讀者，比考慮目的還重要。隨著聽眾和讀者的不同，無論是整體結構、大綱、長度以及用字的難度都會不同。文章的起草人，不論是否刻意，都會考慮目的和對象，這一部分通常在委託他人翻譯時，並不會做充分的說明。相對地，翻譯者即使沒有接到任何說明，也會自行想像目的與對象來進行翻譯作業。不過這畢竟只是想像，不見得與原文起草者心中所想的目的和對象一致。

　　下面介紹的範例是二篇 ABC 市國際少年棒球親善大會的 10 週年賀詞，分別由不同的人翻譯，再經由英文為母語者修改。但是起草者並沒有對翻譯人員或英文為母語者說明這到底是要給從外國來的少棒隊領隊看的呢？還是要給少棒隊的球員看的？而翻譯人員也沒有問及這一點。原文的語氣類似，但是翻出來的英文口氣卻有些微的差異，這是由於翻譯人員或英文為母語者判斷情況上有不同見解的關係。

例文 31　祝賀辭（○）

Congratulations[1] on the 10th Anniversary[2] of the ABC Boys Baseball Association! Your Association's active promotion of this tournament for boys from eight countries, and their homestays in ABC, is a real contribution to international goodwill[3].

I am confident that, for our boys from ABC and for the boys joining us here from all over the world, this exchange of friendship will be a treasured experience[4] that will help them grow to become international citizens in the 21st century. I know that all of us send our support and best wishes for these boys.

In conclusion, may I add my thanks for the efforts and guidance of Mr. Xxxx Xxxxx, President of the Association, and best wishes for the further success of the Boys Baseball Association of ABC.

1) congratulations　恭喜（用複數形，後面加上的介系詞為 on）
2) the 10th anniversary　10 週年　3) goodwill　親善、友好　4) treasured experience　寶貴的經驗（to treasure 是「當作寶貝」的意思）

【內容】

　　ABC 少棒聯盟已經 10 週年了，真是可喜可賀。貴協會積極推動八國少棒賽和 ABC 市的寄宿家庭，可謂真正對國際親善做出貢獻。

　　對 ABC 市和從全球各地前來的球員們而言，這種友情的交流是非常寶貴的經驗，我們也確信這對他們日後成為 21 世紀的國際公民有所助益。我們衷心地支持這些球員們。

　　最後，感謝 Xxxx Xxxxx 協會會長的努力與指導，並預祝

ABC 少棒聯盟今後鴻圖大展。

　　這篇致詞設定的聽眾是 ABC 少棒聯盟的各隊領隊，簡潔有力，很有運動選手的風範。

例文 32　祝賀辭（○）

May I extend my congratulations and heartfelt[1] greetings[2] to you all on this occasion of the International Boys Baseball Tournament being held in this city as a special event to commemorate[3] the tenth anniversary of the ABC Boys Baseball Association.

⋮

To all the players who will participate in this Tournament, may I express my hope that you will use your skills and strength[4] to the fullest[5], and that you will mutually deepen friendships and create much emotion and wonderful memories.

In closing, may I express my heartfelt appreciation to the Boys Baseball Association and to all the other affiliated[6] groups which have done so much to make this Tournament possible, and may I add my own wishes for the success of the Tournament.

---

1) heartfelt　由衷的　　2) greetings　致意（一般用複數形）　　3) to commemorate　紀念　4) skills and strength　技術與能力　5) to the fullest　充分地（一般用 to the full）　6) affiliated　合作

【內容】

　　欣逢 ABC 少棒聯盟 10 週年紀念暨本市舉行國際少棒邀請大賽，在此致上由衷的祝賀之意。

　　　　（中間省略）

　　參加本大賽的各位選手們，希望你們能充分發揮技術和能力，加深友誼，留下感動與美好的回憶。

　　最後，感謝少棒聯盟和努力實現本大賽的所有相關團體，預祝大會成功。

　　這篇短文雖有部分省略，但例文中的每個句子都以 may I... 開頭，是非常客氣的英文，這可能是設定此篇例文為當地來賓的致詞而翻譯出來的吧！

2.　與其回顧過去，不如展望未來

　　一般企業或團體領導人物的新年致詞或社刊中的致詞，往往以回顧過去的國際情勢作為開端。下面的例子是將某協會會刊中理事長的致詞翻譯為英文，翻譯的目的在於對海外的相關團體宣傳。這篇致詞是 1992年寫的，或許已經是舊聞了，但是內容中提到波灣戰爭、莫斯科政變失敗與蘇聯瓦解、泡沫經濟崩潰、日圓升值等主題，所以還是為各位介紹。

例文 33　新年賀辭（原文）

I wish all of the members a happy new year and all the best. Let me recall major affairs during 1991. The Gulf War, started in August 1990 when Iraq invaded Kuwait, threatening a third energy crisis, came to an end in March. The tide for democratization in the USSR and East Europe was accelerated by the failure of a *coup d'état* in Moscow in August, leading to the collapse of the Soviet Union and the birth of the Commonwealth of Independent States (CIS). Substantial changes seem to be taking place in Korea. The talks between south and north may result in a political fusion.

Let's review Japanese economy. Business indicators show that the *Heisei* boom, supported by active investment in plants and equipment and consumer expenditure, has been slowing down since October 1991. The bank rate was decreased at the end of 1991 while the exchange rate of yen showed early in January this year the highest rate during the past three years.

⋮

【內容】

　　祝福各位新年快樂。回顧 1991 年的大事，從 1990 年 8 月伊拉克攻擊科威特肇始的波斯灣戰爭，雖有可能造成第三次石油危機，但總算在 3 月戰爭結束。蘇聯、東歐的民主化風潮因為 8 月莫斯科政變失敗而加速，導致蘇聯瓦解及獨立國協 (CIS) 誕生。在朝鮮半島似乎也逐漸發生重大的變化。透過南北對話，可能會帶來政治的融合。

　　回顧日本經濟，從經濟指標來看，由活絡的設備投資和消費者支出支持的平成景氣，也自 1991 年 10 月以後，呈現衰退的跡象。1991 年底曾調降重貼現率，但是在今年 1 月初，日圓匯率就創下二年來最高的記錄。

【問題點】

　　一般企業或團體的領導者大多喜歡以回顧全球局勢作為新年賀詞的開場白，那麼外國的情況如何呢？看看我手邊社刊的十項要點，各企業社刊中總經理的致詞往往是以國際情勢、國內情勢、業界情勢、自己公司情況等的順序來書寫。而外國公司的總經理致詞則是開宗明義就說到自己公司的狀況，充其量談到公司的外在經濟環境。

　　所以，在閱讀這份新年賀詞的時候，不禁覺得為什麼要將針對國人的致詞翻譯給外國人。以外國讀者為對象的書寫方式，應該也要針對外國人才是。

例文 33 的改善範例（○）

In this first edition of 1992, I would like to wish all of our readers and colleagues[1] every success in your endeavors[2] this year.

In the past year we have seen great changes in our economic and political world. The Gulf War[3] threatened a third energy crisis[4]. Both democracy and confusion were accelerated in the USSR by the failure of the *coup d'état* in August, leading to the collapse[5] of the Soviet Union and the birth of the Commonwealth of Independent States[6]. And substantial change appears about to occur[7] in Korea, with the prospect[8] of political fusion[9].

In the Japanese economy, business indicators[10] showed that the *Heisei* boom, supported by active investment in plants and

equipment[11]) and by consumer expenditure[12]), has been slowing down since 1991. The bank rate[13]) was decreased at the
end of 1991, while in January the exchange rate of the yen[14])
reached its highest point against the dollar in the past three
years.

1) colleague　同事　　2) endeavor　努力　　3) the Gulf War　波斯灣戰
爭　　4) to threaten a third energy crisis　顯現第三次能源危機的跡象。
例如 The black clouds threatened rain.（烏雲表示快要下雨了。）因為
第三次能源危機並沒有真的發生，所以不是 the third energy crisis，
而是 a third energy crisis　　5) collapse　崩潰　　6) the Commonwealth
of Independent States　獨立國協。（原文中在括弧內標示簡寫，表示
下文中還會再次出現；但是事實上只有出現一次，所以就像改善範
例中一樣，不需要列出簡寫。）　　7) about to occur　即將發生　　8)
prospect　預估　　9) fusion　融合　　10) business indicator　景氣指標
11) investment in plants and equipment　設備投資　　12) consumer
expenditure　消費者支出　　13) the bank rate　重貼現率　　14) the
exchange rate of the yen　日圓匯率

【內容】

　　在這 1992年的第一期社刊中，希望各位讀者與同仁今年事
事順遂。

　　去年全球政經有了極大的變動。波斯灣戰爭差點帶來第三次
能源危機。蘇聯的 8月政變失敗也導致了民主化與混亂，造成蘇
聯的瓦解與獨立國協的誕生。在朝鮮半島，由於預期政治即將融
合，似乎也要發生極大的轉變。

　　在經濟方面，由活絡的設備投資與消費者支出支撐的平成景
氣，從經濟指標上來看，從 1991 年起就呈現衰退的跡象。 1991

年年底雖然調降了重貼現率，但是 1 月的日圓匯率還是創下二年來的新高。

【注意事項】

改善範例在不大幅更動原文的範圍內，減少了原文的問題點。描述二年前開始的波斯灣戰爭的用詞很簡單，對蘇聯變革的描述也不難。而平成景氣開始衰退的時期方面，並不像原文中 1991 年 10 月說得那麼詳細，只說 1991 年。

我問過經常閱讀外國企業社刊的朋友，他說外國的領導人往往不太會寫到國際情勢。畢竟股東和員工所期待以及關切的，就是公司的業績以及經營者的展望和方針。

第 Ⅷ 章

提案

1. 利用洗鍊的英文範本

提案是企業在獲取賴以生存的訂單時非常重要的文件，所以在擬定提案時，不僅要提到金額，整篇文章的內容、結構和類型也應該十分注意。看看全球大型企業的提案，從文章的結構、段落的推衍、文章的格調、易讀性、圖表簡單明瞭，甚至在紙質、裝訂等，都花費了許多的心血。

其實，是不是每一份提案都是花了許多時間，鉅細靡遺地擬定的呢？那倒也未必，絕大多數的情況都是趕在繳交期限的最後一秒才完成的。因此，為了解決在短時間內就能顧慮到其中細節的問題，文章就必須標準化。在全球一流企業的電腦內就有好幾種標準的提案段落分段方式，無論是段落或章節，甚至連圖表，都能像模型一樣組裝起來。

所以，要將英文的格式標準化，就應該利用洗鍊的英文範本。我們除了可以參考英美一流企業的提案來學習之外，更可以請英文為母語的人士幫我們檢查。最重要的是，提案的本質在於說服讀者接受筆者的提議，這一點是擬定提案者必須清楚了解的。此外，還需要引起讀者的熱忱和努力，來提高對讀者所產生的效果。

下面介紹的這篇短提案的英文，頗有可議之處。

例文 34　商業提案（原文）

> We, ABC Corporation, are hereby submitting to you the Technical Proposal for the circuit breaker production plant on the basis of the technical meeting held at Singapore on April 18, 199x.

In compliance with your request, we would inform you of the preliminary price for the plant on a basis of the scope of supply and services set forth in the Technical Proposal.

The preliminary price mentioned in this letter is only for the purpose of your budgetary estimation and shall be subject to change according to the Seller's scope of supply and services and other commercial conditions.

The principal terms and conditions on which this preliminary price is based are as follows:

1. The preliminary price for the Equipment to be supplied by the Sellers is:
   $80,000,000 (U.S. Dollar Eighty Million)

2. Ten percent (10%) of the price shall be paid in cash within Thirty (30) days after the effective date of the contract. Ninety percent (90%) of the price shall be paid by means of an irrevocable at sight letter of credit at the delivery of the equipment.

3. The price covers the following Sellers' work and services:
   (a) delivery of the equipment in accordance with FOB Japanese ports, INCOTERMS 1990
   (b) design and engineering service for the equipment to be supplied under the contract
   (c) supply of documents specified in the Technical Proposal

4. The price does not cover the following services:
   (a) services not mentioned in the Technical Proposal
   (b) supervising services for installation and commissioning

5. Other equipment and materials for the plant as well as services not mentioned in the Technical Proposal shall be Buyers' scope.

6. Terms and conditions of the contract shall be confirmed and finalized by discussions between the Buyers and the Sellers before conclusion of the contract.

7. The validity of this price information shall remain until the end of February 199x.

It is very much appreciated if you will give us an opportunity to participate in this project and to serving you.

【問題點】

　　1.本提案第一段是技術提案的附件，第二段以後卻變成商業提案。提案的附件可以分為三種: (1)單純只以寄送為目的; (2)摘要式陳述提案要點; (3)說明商業提案的某部分 (例如有效期限)。上述的例子是商業提案，而非附件。但是第二段卻突然從 In compliance with your request.... 開始，這會讓看了第一段以為是技術提案附件的讀者覺得很突兀。

　　2.這份商業提案不過是為了給客戶估計預算用的預備價格 preliminary price，條件卻規定得非常煩瑣。具體說明如下:

⑴第一、二項的四個數字用阿拉伯數字和拼字兩種寫法來表示慎重,但是有點不自然,只用阿拉伯數字就足夠了。

⑵第二項的付款條件中提到在簽約生效後 30 天之內支付 10% 的訂金,但是此時並不明確說明 30 天的期限;irrevocable at sight letter of credit(看過後即不可取消之信用狀)在這裡也只要用 letter of credit(信用狀)就夠了。

⑶第三項 FOB(free on board,船上交貨)明確的定義是根據 Incoterms(國際商工協議所貿易用語定義) 1990 版。

⑷第三項中說明估價包含的範圍,這是有必要的。但是第四、五項估價範圍外的規定,對估計預算而言,實在沒有必要,因為第三項中說明的範圍以外當然指的是 ABC 公司的估算範圍之外而言。

⑸第七項的有效期限 (validity) 因為表明不可能接受這個 preliminary price 後下單,所以實際上是不需要的。但是沒有說明價格的有效期限,彼此也會擔心,所以這樣的寫法也不是不可以。或者也可以換一個說法,比方說: This price is valid until the end of February 199x, subject to our confirmation. (此價格到199x年2月底前有效,但請與我們確認。) 這表示價格在 2 月底前是否有效,就算對方願意接受報價,還不等於合約成立,必須再與我方確認。

例文 34 的改善範例(○)

① ABC Corporation is pleased to enclose our Technical Proposal[1] for a circuit-breaker manufacturing plant on the basis of the technical meeting held between us in Singapore on April 18, 199x.

② In addition, as you requested, we here offer a preliminary price, based on the scope of supply and services set forth[2] in the Technical Proposal. This preliminary price is only for the purpose of your budgetary[3] estimation, and is subject to[4] change according to the scope of supply and services and other commercial conditions.

③ Preliminary price terms and conditions[5] are as follows:

1. The preliminary price for the equipment to be supplied by ABC Corporation is $80,000,000.

2. Ten percent of the price shall be paid in cash upon effectuation[6] of the contract. Ninety percent of the price shall be paid by means of letter of credit upon the delivery of the equipment.

3. The preliminary price covers the following by ABC:
a) delivery of the equipment FOB Japanese port, Incoterms 1990
b) design and engineering for the equipment to be supplied; and
c) supply of documents specified in the Technical Proposal.

4. This preliminary price will remain valid until February 28, 199x.

④ We very much look forward to the opportunity to participate in this project and appreciate your consideration.

1) Technical Proposal 技術提案（書）（提案一般是由 Technical Proposal, Commercial Proposal 和 Management Proposal 三個部分組成。
Management Proposal 是提案者說明此一計畫如何進行的部分）
2) to set forth 陳述、說明 3) budgetary 預算 (budget) 用的
4) subject to 以～作為條件 5) price terms and conditions 價格條
件（terms 和 conditions 是相同的意思） 6) effectuation （合約）
生效

【段落大綱】

① 說明附上技術提案的根據。
② 說明下列預估金額的性質（預定價格）與目的（預算用）。
③ 說明預估金額的條件：
   1. 預備報價的金額
   2. 付款條件
   3. 估價範圍包含項目
   4. 預估金額的有效期限
④ 表達積極的態度並作結。

【內容】

　　ABC 公司依據 4 月 18 日新加坡技術會議，提出電路阻斷
器生產工廠的技術提案書。

　　此外，依照您的要求，依據此技術提案書中談及之交貨和服
務範圍，提出預估價格。此預估價格乃供您預估預算之用，隨交
貨、服務範圍及其它交易條件不同，得加以變更。

預估價格的條件如下：

1. ABC 公司交貨的機器預估價格為 8,000 萬美金。

2. 此價格的 10%，於合約生效時以現金支付。其餘 90% 則在機器交貨後以 L/C 支付。

3. ABC 公司的預估價格範圍如下：

a) FOB 日本港口 (Incoterm 1990) 的機器交貨。

b) 交貨機器的設計與工程。

c) 提供技術提案書中所述的文件。

4. 此預估價格於 199x 年 2 月 28 日前有效。

期待有機會能參與本計畫，並懇請考慮。

【注意事項】

來看看原文與改善範例的差別。

1. 原文中有七項條件，但是在改善範例中只有四項，這是因為在原文中第四、五項說明不包括在預估範圍內的項目；而改善範例中省略了這一部分，因為不包括在預估範圍內的項目（改善範例第三項），就不在報價的範圍內，這種邏輯是非常明確的。

2. 原文第一行 We, ABC Corporation, are...這種用法現在已經很少人用了，反而是不用逗號 We at ABC Corporation are...的情況比較多。此外，像是 We are submitting... 或 ABC Corporation is submitting... 也可以。在改善範例中用的是 ABC Corporation is pleased to enclose...。

3. 原文第一句的 at Singapore 應該是 in Singapore。

4. 原文第二段突然從 preliminary price 的新話題開始，難免讀者吃驚。改善範例的第二段則是用 In addition, as you requested, 圓滑地將話題從技術提案轉換為商業提案。

5. 改善範例中第一項的金額只以阿拉伯數字標示。而第二項的 Ten percent、Ninety percent 不寫成 10% 和 90%，是依據句子如果以數字起頭，則避免使用阿拉伯數字的一般原則。

（參考）　**shall**的用法

　　改善範例中 shall 的用法如第二項的 shall be paid，意思是「應被支付」或「必須支付」。除了 shall be paid 的用法之外，must be paid, should be paid 或 is to be paid 的用法可行嗎？為了解開這個疑問，我們來看看文法的解釋。

　　第一人稱加上shall，表示「單純未來」（不代表說話者的意志）。

I **shall** be 40 years next month.

（下個月，我就 40 歲了。）

　　相對地，第一人稱如果加上 will，就代表說話者的意志。

I will wait for you. (= I intend to wait for you.)

　　但是，單純未來的 shall 在目前除了非常形式化的場合之外，都不會使用，而是以 will 代替。

I will be 40 years next month.

　　但是在疑問句中，現在還是有人用 shall。

Let's go, **shall** we?

（走吧？）

What **shall** I do with this letter?

（這封信我該怎麼辦？）

　　另外在表示決心時，也可以用 shall。

I **shall** return.

（太平洋戰爭期間，麥克阿瑟將軍在遭到日軍攻擊而離開菲律賓時的名言。）

　　第二人稱或第三人稱加上 shall 時，表示說話者（第一人稱）的意志。

You **shall** have a sweet. (= I'll give you a sweet.)

He **shan't** come here. (= I won't let him come here.)

但是 you shall 有強迫對方接受自我意志的感覺，所以上述的例子只用於針對小孩或寵物。

第三人稱加上 shall 就變成命令。這種用法常見於法律、規定、規格、合約和規格書中。

・**運動規則中，**

Yachts **shall** go round the course, passing the marks in the correct order. [Yacht Racing Rules]

（帆船應依正確順序通過標示，繞行一周。）

・**憲法中，**

Article 10. The conditions necessary for being a Japanese national **shall** be determined by law.

（第 10 條，身為日本國民的條件，由法律規定之。）

・**合約中，**

The Seller **shall** deliver the goods to the Buyer within thirty days after the effective date of this agreement.

（賣方必須在合約生效起的 30 天內將商品交給買方。）

・**規格書中，**

Window sashes **shall** be primed before glazing.

（窗子的外框必須在裝上玻璃前先塗漆。）

前面改善範例中的 shall 就是「合約書中的 shall」。提案雖然不是合約，但是在實務書信中提到雙方或對方（買方）的義務時，習慣以與合約相同的方式來使用 shall。舉例而言：

The quantity of all materials to be supplied **shall** be decided upon completion of all installation drawings.

（應供應之所有材料數量，應於所有裝置圖面完成後決定。）

This proposal is not an offer and acceptance of it by Buyer **shall** not constitute a contract between the parties.

（本提案並非報價，買方不得依此報價視為兩者間之合約。）

但是提案的本質在於向對方提出建議，所以基本上不應該用 shall 而應該用 should。

We propose that this project **should** be implemented imme-diately.

（我們建議此計畫應立即執行。）

此時也可以用沒有 should 的 We propose this project be implemented immediately. 這種不用 should 的方式比較普遍。

如上所述，在法律、合約及規格書中常用的 shall 擁有法律上的強制效力，意思與 must 的「非～不可」相同。相對地，should 是「應～」，是沒有強制力的提議。在商務書信中也常用 to be。

You **are to** stay here.

（你得留在這裡。）

The agreement **is to** be signed next week.

（合約將在下週簽字。）

由於 to be 可以用在強制和預定，所以意思並不很明確。

接下來，將表示強制和義務的助動詞整理如下：

shall	（必須）有法律的強制效力。
must	（必須）無法律的強制效力。
have to	（必須）外界給予的義務。
should	（應該）建議，表示道德上的義務。
ought to	與 should 一樣，但是在商業書信中不太使用。
had better	（～比較好）表示建議，有時表示威脅。
can	（可以～）表示能力、可能性或許可。
may	（可以～）表示許可或不確實的可能性。

2. 建議、請求或要求

　　所謂的對外提案 (external proposal) 是對解決對方問題而作的企劃（對公司內部的提案叫做 internal proposal），為說服對方採用的行為。請求雖然也是說服對方的行為，但是它的目的不在於解決對方的問題，而是解決自己的問題。而要求則是以更強於請求的態度來面對對方。順帶一提，英文 claim 的意思是針對自己的所有物，或是指要求自己的權利而言 (to demand as one's own or as one's right, *Black's Law Dictionary*)。

　　下面的例子是提議客戶變更設計的傳真。

例文 35　建議變更規格的傳真（原文）

As specifically indicated in your specification, particular consideration should be taken into account for the design of the outlet channel configuration of the boiler, because it is important for this item to prevent undesirable phenomena like the accumulation of heavier components inside the outlet channel, etc.

We also have recognized the importance of the consideration for the part, and would like to propose the adoption of the configuration different from your original idea.

We would very much appreciate your review over our consideration and like you to give us your approval of our design by the first week of February.

【內容】

　　誠如貴公司規格書中所示，鍋爐出口管的形狀設計必須要特別考量。因為對這個設備而言，不讓沈重的成份累積於出口管上以防止不良現象發生，是非常重要的。

　　我們也了解考量這部分的重要性，所以建議採用不同於您原案的形狀。

　　請研究我們的看法，並懇請於2月的第一個禮拜前同意我們的設計。

【問題點】

　　這份傳真基本上是提案，含有請求和要求的要素。第二段中雖然同意客戶認為出口管形狀的重要性，但又提議與對方腹案迴異的形狀。由於這兩種看法用 and 連接，會讓人覺得很奇怪。

　　第二段是提案，第三段卻變成 We would very much appreciate 的請求口吻，之後又變成 like you to give us your approval 的要求語氣。

　　接下來，在這些文字中出現三次的 consideration，用法上有些問題。第一段 consideration 的用法是 particular consideration should be taken into account, take something into account = to give proper consideration to something (*LDCE*)，所以這句話就變成 to give proper consideration to particular consideration，等於意義重複。第二段 the importance of the consideration 中的 consideration 是 careful thought（考慮）的意思，與第一段的 consideration 不同。第三段的 our consideration 還不如 our idea 來得容易懂。同樣一個 consideration，却有不同的意義，這樣會造成讀者的困擾。

　　第一段結束的 etc. 可能是「等」的概念，但是 etc. 語焉不詳，商務書信中最好不要用。尤其這裡只舉出 the outlet channel 一個例子，就該避免使用 etc.。改善範例中沒有使用 etc.，如果

一定要用「等等」的話，可以用 such as 或 for example。這裡的 the outlet channel and other components 就是一個好例子。

下面的改善範例就是以要求的語氣來寫的。

例文 35 的改善範例（○）

① As has been shown both by our own investigation and by the information in your specification, a change in the configuration of the outlet channel for the boiler is required. Following are notes and figures indicating design changes necessary, along with reasons for them.

② May we ask that you review these notes? In order that the project proceed on schedule[1], we will have to reach an agreement[2] on these design changes by February 5th.

③ As you will see within, our main concern is to avoid accumulation of heavy components inside the outlet channel. We believe that specification also points to[3] this problem.

④ We look forward to hearing from you soon on these changes, and to working with you on this project.

1) to proceed on schedule 依計畫進行　　2) to reach an agreement 達成協議、共識　　3) to point to 指～

【段落大鋼】

① 說明變更規格的必要性，介紹例文以後的說明資料。
② 請求研究變更規格，提示回答日期。
③ 說明變更的主要目的。
④ 催促回答，表達對協助的期望，並作結。

【內容】

如本公司調查及貴公司的規格書資料所示，此鍋爐出口管的形狀必須變更。以下為設計變更時所需事項的簡單說明、圖面及其理由。

懇請研究此一簡單說明。為了讓計畫如期進行，必須在 2 月 5 日前獲得變更設計的同意。

如本文中所述，我們注意到不讓沈重的成份累積於出口管上。我想您的規格書中也指出了這個問題。

請就此變更事宜回答。

【注意事項】

改善範例是以要求的口氣來寫的。看看第二段 May we ask that you review these notes? 感覺很有禮貌，但是 In order that... we will have to reach an agreement on these design changes by February 5th. 就不是請求，而是直接的要求。那要如何讓這句話變得客氣一點呢? 我們可以這樣說，In order that... we hope to be able to reach an agreement on these design changes with you by February 5th.

另外一個值得注意的地方是改善範例的邏輯結構。原文是先說明變更設計的理由，接下來是提議變更設計，最後是要求對方的行動。改善範例則是先說明變更設計的必要性，再談到要求對

方的行動，最後才是設計變更的理由。我們在下一個例子中來看
看其間的差異。

3. 從整體到細節

在商業書信中，如何讓讀者在短時間內掌握內容，是非常重
要的。為了達到這個目標，最好是依據問題點、原因、摘要、結
論等，從整體開始下筆，詳情則放在總論之後。這種原則稱為從
整體到細節 (general to particular)，或者叫做演繹法順序 (deductive order)。所謂演繹法 (deduction) 指的是從一般的原理中
導出特殊原理和事實的推論方法。相反的推論方法，也就是從各
種事實中導出一般原理的方法，稱為歸納法 (induction)。就算是
以歸納法來推論，只要將報告的推論順序倒過來，從總論的整體
部分開始寫，同樣也是 general to particular。這種原則可說是商
業書信的各種原則中最常為人使用的。

接下來介紹的傳真信，是建築商 ABC 公司在計算海外客戶
XYZ 公司的工程成本時，發覺超出當初的預算，因此建議 XYZ
公司縮小工程範圍。

例文 36 建議縮小工程的傳真 (原文)

199x July 28

REF. No. B91–6–7

FAX to: Mr. Le Berre

　　　　Mr. Belsta　　　　　　　From: ABC Co., Ltd.

CC: Mr. Lopez→ Mr. Megemont　E. Taylor

　　　　　　　　　　　　　　　Overseas Planning Dept.

Re: Land and Building Budget

Dear sir,

Thank you for your fax (July 26).

On the base of[1] your information about the construction site, we have estimated the actual cost and compared[2] with the original budget. (See Page 2/9)

To keep the cost within the framework of the budget, we made counter step[3] for cost reduction as shown on the page 2/9.

Please inform us of your opinions whether you can agree about the following items[4] in the counter steps by August 9[5].

(1) Land area shall[6] be reduced to 50,000 m^2 (attached sheet I, VI, VII[7])

(2) Plant area shall[6] be reduced from 8,000m^2 to 7,200m^2 (120×60 m) (Our engineer staffs, MR. SMITH, MR. OWEN, MR. FORD[8] already discussed the plant reduction plan with MR. LOPEZ and MR. MEGEMONT in MADRID at the end of May.)

(TO MR. MEGEMONT)
Please inform us how far XYZ can follow up the cost reduction plan itemized among us in ROME by August 9[9]. (See attached sheet III, IV, V[10])

Best Regards,

【內容】

我們依據您提供關於建築工地的訊息來預估實際的成本，並與原來的預算做了比較。（參照 2/9 頁）

為了讓成本維持在預算之內，我們提議如 2/9 頁的成本削減方案。

請於 8 月 9 日前通知是否同意下列的方案項目。

(1)減少建設面積至 5 萬平方公尺（參照附件Ⅰ，Ⅵ，Ⅶ）。

(2)工廠面積 8 千平方公尺中減為 7 千 2 百平方公尺。

　　（本公司工程師史密斯先生、歐文先生和福特先生已經與
　　洛派茨先生、梅格蒙特先生於 5 月底在馬德里討論過工
　　廠縮小案。）

　　（致梅格蒙特先生）

請於 8 月 9 日前告知我們在羅馬提及項目的成本削減案，XYZ 公司能夠遵從到什麼程度。（參照附件Ⅲ、Ⅳ、Ⅴ）

【注意事項】

1.先從 general to particular 的觀點來看，由於所有的詳細資料都包含在附件中，所以在這篇短文中只需提到要點即可。但是下面這一點也讓人覺得很奇怪。

在這份傳真中，最重要的內容似乎是最後對梅格蒙特先生的請求，但梅格蒙特先生卻只是副本的受文者，正本的受文者另有其人。所以傳真的受文者恐怕會躊躇，而不知所措。

另外，標題只有 Land and Building Budget，受文者無法預測到底發文者希望他採取什麼行動。一般是僅看標題就能了解發文者到底想要傳達什麼事情及要求什麼。

2.傳真的受文者似乎是勒貝爾先生和貝爾斯塔先生，但在開頭的稱謂卻是單數的 Dear sir。而且 sir 也不正確，S 必須大寫。CC 到底是發文者 ABC 公司一開始就發了二份傳真呢? 還是希望受文者影印後分發? 也沒有交代清楚。發文者如果發的不是正

本而是影本的話，在上面寫 Copy by sender 這一類的標示語，受文者也比較放心。如果希望受文者影印後發送，就寫明 Copy by receiver 或 Please distribute copy to～，那麼受文者應該就會影印後分發好。

3.依照原文標示的號碼，逐一仔細地來研究研究。

1) on the base of 是錯的，on the basis of 才是正確的。

2) 因為 compare 是及物動詞，所以應寫成 compared it with。

3) we made counter step 不是「擬定方案對策」，而是已經執行對策的意思，所以最好用 we propose。

4) agree about the following items 是「就下列項目而言，意見一致」；如果是「贊成下列項目」的話，應該是 agree to the following items。

5) by August 9 在這個位置就等於修飾 agree，會被解釋為「能否在 8 月 9 日前贊成同意」。如果要修飾 inform，則應該寫成 Please inform by August 9。

6) 因為這個 shall 是合約或法律中所使用的 shall，對提案而言，語氣太強了些。第(2)項中的 already discussed 會被解釋為已經決定好了。

7) attached sheet Ⅰ, Ⅵ, Ⅶ 應該寫成 attached sheet Ⅰ, Ⅵ and Ⅶ。

8) MR. OWEN 與 MR. FORD 之間應加上 and。此外，staff 表示全體員工，雖然是複數形，但不用 staffs。再加上這份傳真中的人名都大寫，實在很礙眼。

9) 這個 by August 9 的位置等於是在修飾 follow up，必須放在修飾 inform 的位置。

10) 參照上述 7)。

例文 36 的改善範例 A（請求研究提案）

To : Mr. Le Berre and Mr. Belsta

From: E. Taylor

Copy by sender[1]: Mr. Lopez and Mr. Megemont

Re : Our Proposed Changes in Land and Plant Areas

Dear Sirs:

Based on the information we received from you regarding the plant site and the costs associated with it, we propose the following cost reduction[2] measures (as discussed[3] between Mr. Smith, Mr. Owen, and Mr. Ford of our engineering staff, and Mr. Lopez and Mr. Megemont of your office at the end of May) in order to keep costs within the budget for this project:

1. Reduction of land area from 60,000 m^2 to 50,000 m^2
2. Reduction of plant area from 8,000 m^2 to 7,200 m^2

I would like to ask for your responses to these changes by August 9, so that preliminary drawing can begin on schedule.

Best regards,

1) copy by sender　表示此份傳真的影本由發文者分發給下列人員
2) cost reduction　降低成本，英文中沒有 cost down 的說法

【內容】
　　主旨：建廠用地與工廠面積變更的提案
　　依據您提供的工廠工地及相關的成本訊息，為了讓成本維持在計畫的預算額度之內，我們提議下列的成本削減方法。（內容與本公司工程師史密斯先生、歐文先生、福特先生與貴公司的洛派茨先生、梅格蒙特先生在 5 月底協商相同。）
　　　1.將用地面積由 6 萬平方公尺縮小為 5 萬平方公尺
　　　2.將工廠面積由 8 千平方公尺縮小為 7 千 2 百平方公尺
　　為了讓預備圖面的製作能夠如期開始，請於 8 月9 日前針對此一變更做出回答。

【注意事項】
　　改善範例 A 是為了削減成本的提案。
　　1) 的 copy by sender 表示發文者會影印後發送，如果要受文者影印後發送，可以寫成 copy by receiver 或 Please copy and distribute to...。
　　3) as discussed...是把客戶已經答應這件事儘早告知對方。
　　第一項的用地面積中加上「由 6 萬平方公尺」，和第二項的工廠面積保持對等。
　　最後一句說明設定回答期限的理由。

例文 36 的改善範例 B（要求對方認可己方的決定）（○）

Based on the information we received from you regarding the

plant site and the costs associated with it, we will implement the following cost reduction measures (as discussed between Mr. Smith, Mr. Owen, and Mr. Ford of our engineering staff, and Mr. Lopez and Mr. Megemont of your office at the end of May) in order to keep costs within the budget for this project:

1. Reduction of land area from 60,000 m^2 to 50,000 m^2
2. Reduction of plant area from 8,000 m^2 to 7,200 m^2

Please let us know by August 9 what these changes will require from your side. We intend to begin preliminary drawings before the end of the month.

【內容】

　　基於我們從您所收到關於工廠工地和相關成本的訊息,為了讓成本維持在預算內,我們將執行下列的成本削減方案(與本公司工程師史密斯先生、歐文先生、福特先生和貴公司洛派茨先生、梅格蒙特先生於 5 月底協商的內容一致。)

　　1.將用地面積 6 萬平方公尺縮小為 5 萬平方公尺

　　2.將工廠面積由 8 千平方公尺縮小為 7 千 2 百平方公尺

　　這項變更的結果,您有什麼需要,請於 8 月 9 日前告知。我們預定於月底開始製作預備圖面。

【注意事項】

　　跟改善範例 A 的差異在於斬釘截鐵地說明「本公司將執行下列成本削減方案」(第一句),以及「變更的結果,您有什麼

需要，請於 8 月 9 日前告知」（倒數第二句），沒有給對方選擇的餘地。如果已經確定要採用此方案，而要通知這項既成事實時，口氣就會是這種感覺。但是對客戶而言，這樣的語氣不太禮貌。

例文 36 的改善範例 C（在一開頭就說明要求對方的行動）（○）

I would like to ask you that you let me know by August 9 any changes that will be necessary in your planning as a result of the following changes that will be necessary at the plant site in order to keep the costs within the budget for that project:

1. Reduction of land area from 60,000 m² to 50,000 m²
2. Reduction of plant area from 8,000 m² to 7,200 m²

Note that these are the same changes as discussed between Mr. Smith, Mr. Owen, and Mr. Ford of our engineering staff, and Mr. Lopez and Mr. Megemont of your office at the end of May.

【內容】

　　為了讓成本維持在本計畫的預算之內，工廠工地必須有以下的變更，請於 8 月 9 日之前，告知貴公司在變更計畫中所需的變更點。

　　（中間省略）

　　這些變更的內容，與本公司工程師史密斯先生、歐文先生、福特先生和貴公司洛派茨先生、梅格蒙特先生於 5 月底協商的內容相同。

【注意事項】

　　首先說明對 XYZ 公司的要求，之後再說到原因，這也是 general to particular 的順序。

（參考）各種順序

　　在商業書信中，光是將每一個訊息做正確而易懂的傳達是不夠的。清楚呈現每一件事實或構思的相互關係以及其在文中的定位，將可使文中的訊息變得有價值，同時引起讀者的反應。為了達到這個目標，訊息必須要有一定的組裝順序。在這種順序之中，最普遍的就是「從整體到細節」(general to particular) 的順序。

　　接下來說明包括 general-to-particular order 在內的幾種基本順序。

演繹法順序 (deductive order = general-to-particular order)

　　這個順序與「從整體到細節」相同，和下面的歸納法成對比，我們稱為演繹法順序。商務書信，尤其是希望讀者在短時間內就能正確掌握內容的報告書，其大原則就是演繹法順序。在文章開頭就先寫下要點、結論、重要建議事項，而支持的詳情和說明則寫在其後。如果還有更詳細的資料，則作為附件。演繹法順序不僅是文章結構完整的原則，同時也是商業書信和段落結構的原則。

　　所謂的演繹法，就是從一般的原理中推論特殊原理或事實的方法，但是要讓文章的結構有演繹法的形態，推論的方法卻不是演繹法。一般而言，大多會以歸納法，也就是從幾個具體事實中引出一般的原理。思考是歸納法式的，但是在報告的時候，就把順序顛倒，以演繹法來陳述。所以文章的開頭都是在正文寫完之後才下筆的。

演繹法順序的好處在於能讓讀者在短時間內了解要點、盡快判斷；不需要的話，也用不著整篇都看完。缺點則是如果讀者的看法與結論相反，一開始就談結論，恐怕會造成讀者的反感。

歸納法順序 (inductive order = particular-to-general order)

與演繹法的順序相反，先舉出事實，然後從事實中導出結論。歸納法順序的優點，在於可以改變已經有先入為主觀念的讀者看法，但是在到達要點之前，必須要看好多的事實和議論，會讓讀者不耐煩。

重要性順序 (order of importance)

這是從最重要的項目逐漸記述到重要性較低項目的一種方法。重要性的判斷很主觀，所以要特別小心，例如對依重要程度列出客戶清單的廠商而言，客戶如果覺得競爭廠商的重要性竟然列在自己公司之前，一定非常惱火。

也有從比較不重要事項開始推到重要事項的做法，這時必須讓讀者知道，最後面的項目才是最重要的。例如，要先給一個預告，像是 "Finally, and most important..."。

時間順序 (chronological order)

依時代或時間的先後記述。時間順序經常用於流程說明或程序的指示，這種順序也稱為 step-by-step order。

空間順序 (spatial order)

機器的功能或結構零件的說明等，會依照空間的順序。例如往右、由上到下、由北到南等。

字母順序 (alphabetical order)

用於列舉國名或公司名時。例如在列舉國名時， the U.S.A., the U.K., Germany, France, Canada, Australia, Korea, and the Phillipens，這種排列方式，會讓讀者以為是有什麼重要的原因

才如此排列。如果以字母順序來排列，就不會造成誤會。也就是說，應該排列為 Australia, Canada, France, Germany, Japan, Korea, the Phillipines, the U.K., and the U.S.A.

4. 有深度的內容，就要用有深度的英文來表達

有許多人從國中、高中到大學一直學習英文，即使進入社會之後，也在公司內上語言課程來學習英語會話，或是訂閱英文報紙、雜誌，保持對英文的興趣。但有很多人認為自己的英文程度並不好。

其實我們無須那麼看輕自己的英文程度。雖然發音破、又不流暢，但還是努力地說寫有深度的英文，很多外國人也都讚許這一點。國際企業的員工不僅要會英語會話，還要能用英語討論，用英語談判。這種能力的基礎之一，就是能夠正確地書寫有深度的內容。下面的例子，就在傳遞有深度的內容。

例文 37　建議不要變更交易條件（原文）

> ATTN : (XYZ) Mr. Peter ROGERS　　　　　19 Feb. 199x
> FROM: (PR) S. SCOTT
> REF　: Article 3 of Sales Agreement: Delivery and Risk of
> 　　　　Loss
>
> Here I would like to arrange the following point described in your fax XYZ–1772.
>
> You mentioned the trade terms between XYZ and ABC should be "C&I Liverpool". On the other hand, the trade terms on

Quotations and Purchase Orders have been "C&I U.K." for these years.

From the viewpoint of range of cargo insurance, "C&I Liverpool" is more suitable as you said. But "C&I United Kingdom" is also correct and there is no problem and no demerit for both of XYZ and ABC with this terms, because this Article 3 says the range of cargo insurance.

Furthermore, if the terms are corrected to C&I Liverpool, it will cost much labor, namely, amendment of master registration of the terms, change of all of quotations and purchase order sheets, and notification to related divisions and sections in XYZ. So we would like to leave it as "C&I United Kingdom."

Would you please let us know your opinion about this matter as soon as possible?

That is all.

Best Regards,

【內容】
　　（大意）XYZ 公司的羅傑先生指出，XYZ 公司與ABC 公司之間的交易條件，不是 C&I UK，而是 C&I Liverpool。針對這一點，ABC 公司的史考特先生說 C&I UK 指的是海上保險的

範圍，這種說法也正確。如果現在要更改交易條件的登記記錄，就連報價單和訂單等都得更動，勞師動眾，所以建議就不要再更改了。

【問題點】

　　只要了解這裡所使用的貿易用語，就能夠了解這份傳真在說些什麼。受文的羅傑先生當然也應該懂得這份傳真在寫些什麼，如果要指出比較明顯的問題有三：

　　1.第一句 arrange 這個動詞不適合這裡使用。

　　2.最後的 That is all. 可能是「以上」的直譯，當然外國人不這麼用。

　　3.一開始的 19 Feb. 199x 不要簡寫比較好，最好還是寫成 19 February 199x。

例文 37 的改善範例 (○)

---

19 February 199x

To　 : Mr. Peter Rogers
　　　　XYZ
From: Sam Scott
　　　　Procurement Department
Ref　: Your fax XYZ1772

Re　 : Article 3 of Sales Agreement—Designation of Trade
　　　　Terms: Suggestion to Continue Present Practice

① We would like to suggest and request that trade terms

---

continue to be designated "C&I United Kingdom", as in the past.

② As you have said, the designation you suggest, "C&I Liverpool" is an accurate term for the range of cargo insurance; however, since cargo insurance is described elsewhere in Article 3, we believe the present designation is acceptable both for XYZ and for ABC.

③ The problem is that changing the terms is very costly in labor, namely in amending master registration of the terms, changing all quotations and purchase sheets, and giving notice of the change to all related divisions and sections of XYZ and ABC. If this is necessary or beneficial, we will, of course, do so; but if not we prefer to continue the present practice.

④ Could you let us know if you think this will be possible?

【問題點】

　　這裡所用的 terms 是「條件」的意思，也就是 terms of delivery（交貨條件）。C&I United Kingdom 是指貨到英國為止、含保險費在內的價錢，也就是在裝貨船港將貨物裝到船上時的裝船價格加上到達目的地港口的海上保險費用。

【段落大鋼】

---

① 說明我方提案要點。
② 肯定對方的提案，更主張我方提案的正確性。
③ 指出對方提案的問題點（成本較高）。
④ 要求對方同意本方案。

【內容】

　　主旨：銷售合約第三項：持續現行交易條件名稱的提案
　　我們建議及提議交易條件的名稱維持以往的 C&I UK。

　　誠如您所說的，您提議的 C&I Liverpool 是表示貨物保險的正確用語；但是我們相信貨物保險在第三項中某處也有提及，所以現行的名稱對 XYZ 公司及 ABC 公司而言都是可以接受的。

　　問題在於變更條件時手續繁雜。換句話說，交貨條件登記的變更、所有估價單與訂單的變更，以及通知所有 XYZ 公司與 ABC 公司相關部門變更事宜等，非常麻煩。如果變更是有必要或有利的話，當然應該變更；如果不是如此，希望能維持現行的方式。

　　請通知可行與否。

【注意事項】

　　改善範例也是採用 general to particular 的順序，一開始就做提案，之後再說明理由。
　　標題雖然長了一點，但是將傳真的內容說得很具體。

5. 表達方式應注意禮節

　　書寫時除了避免很粗野或失禮的英文之外，還要站在對方的立場考慮，以謙虛的態度來書寫。

例文 38 建議變更受訓行程的傳真（原文）

Re: TRAINING SCHEDULE

We received your fax (dated Jan. 15th, Fax No. XYZ/F011).
We do not agree with your suggestion about 2/25–3/18 training
schedule.
So we propose a new plan for training from 18th (Mon) to
25th (Mon).
(Refer to next page)
Please reply your opinion.

Best regards,

P.S.
A map to your hotel (…)

【問題點】

第二句突然說到 We do not agree with your suggestion 就有
點不客氣，而且沒有說明反對的理由，更顯得失禮。我們曾經在
第Ⅳ章第 3 節中提到，說明理由也是尊重對方的一種禮貌表現。
結尾的 P.S.（postscript: 再啟）也不正確。

例文 38 的改善範例（○）

Re: Alternative[1] to Your Proposal for Training Schedule,
Your Fax No. XYZ/F011 of January 15

Unfortunately, your proposal for a training schedule from February 25 to March 18 is impossible because of prior commitments[2] in the Training Center. We propose an alternative, February 18 (Monday) to February 25 (Monday). Details are on the following page. Could you let us know by the end of the week if this schedule is possible from your side, so that we can reserve training facilities and make other preparation?

Also accompanying this message is an explanation of transportation from your hotel to our training center.

Best regards,

1) alternative　替代方案　2) prior commitment　先前的約定

【內容】

主旨：您所提議受訓行程的替代方案

您 1 月 15 日的傳真XYZ/F011

很遺憾，您建議 2 月25 日到3 月 18 日的受訓行程，由於受訓中心事前已有其它安排，所以無法使用。我們提議 2 月 18 日（星期一）到 2 月 25 日（星期一）的替代方案。詳情列於下頁。我們希望能預約受訓中心以及進行其它準備工作，所以請於週末之前，回覆您認為此行程是否可行。

茲附上從旅館到受訓中心的交通說明。

第 IX 章

簡介

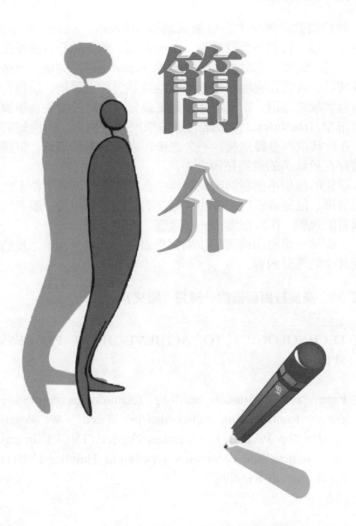

1. 引起讀者爭論

我們與歐美傳統中對辯論或議論 (dispute, argument) 的觀念很不一樣。我們不喜歡議論，也不多話，認為能心有靈犀是美德。但是，歐美文化從古希臘的廣場 (agora) 辯論開始，便擁有這種傳統。人們在廣場上毫無隱藏地談論，相互爭辯，這種習慣使得他們能言善道，希望語言能夠更具效果，終於衍生為辯論術與修辭學 (rhetoric)。古希臘的修辭學歷經古羅馬、中世紀的文明，在近代歐洲登峰造極。不久之後，古典修辭學衰退，但精髓仍舊存在於歐美的溝通技術之中。

辯論的技術不僅在演講或談判中，在文章的溝通工作上也能大有發揮，這是筆者對讀者的挑戰，引起讀者的爭論。讀者也歡迎筆者的挑戰，作好準備，一面回應一面往下讀。

下面是一份影印機商品技術報告書（公司內外用），我們來研究其中的部分內容。

例文 39 產品技術報告的一部分（原文）

> ## 4. TECHNOLOGIES TO ACHIEVE HIGH PROCESS SPEED
>
> Paper and document handling technologies contribute greatly to increasing copier process speed. We developed a Top Vacuum Corrugation Feeder (TVCF) for paper handling and a Vacuum Document Handler (VDH) for document handling.

4.1 TOP VACUUM CORRUGATION FEEDER (TVCF)

This section compares the TVCF with other paper handling methods and explains the TVCF feed cycle.

4.1.1 Comparison of TVCF with Other Paper Handling Methods

The conventional paper handling methods are:

⋮

【內容】

　4.高速處理技術

送紙與文件傳送技術對增快影印機的處理速度有極大的貢獻。本公司在送紙方面開發了 Top Vacuum Corrugation Feeder (TVCF),而為文件傳送開發了Vacuum Document Handler (VDH)。

　4.1 Top Vacuum Corrugation Feeder (TVCF)

本節說明 TVCF 與其它方式的比較情況以及送紙循環。

　4.1.1 TVCF 與其它送紙方式的比較

以往的送紙技術……（以下省略）

【問題點】

　這份報告書的開頭客觀地說明新技術，但沒有提到新技術是否較為優異。如果這份文件是公司內部用的話，那也就算了；但如果是針對公司外部的話，就必須藉著類似改善範例的方式，挑起讀者的爭論。

　就英文方面的問題而言，第一句和第二句之間沒有邏輯關

連，實在讓人覺得很奇怪。只要寫成 Since paper and document handling technologies contribute..., we developed... 就能產生邏輯關連。

例文 39 的改善範例（○）

---

4. TECHNOLOGIES[1] TO ACHIEVE HIGHER COPYING SPEED

Present paper and document handling technologies do not allow an increase in copier process speed. Our new Top Vacuum Corrugation Feeder (TVCF) for paper handling, and Vacuum Document Handler (VDH) for document handling overcome the slowness of present methods.

4.1 TOP VACUUM CORRUGATION FEEDER (TVCF)

For high speed copying, paper handling devices must be compatible with[2] high speed technologies and be capable of handling all common weights of paper without creating paper debris[3]. This section compares TVCF with slower, less flexible and less reliable paper handling methods, and explains the TVCF feed cycle.

4.1.1 Comparison of TVCF with Other Paper Handling Methods

The conventional paper handling methods are:

---

1) technologies　工業技術。這裡指的是眾多領域的技術，所以用複

數。類似的例子還有 The system uses advanced computer and satellite technologies. (*LDCE*)　2) compatible with　相容於～　3) debris 碎片

【內容】

4.高速處理技術

目前的送紙與文件傳送技術，無法增加影印機的處理速度。本公司新的送紙技術 Top Vacuum Corrugation Feeder (TVCF) 和文件傳送技術 Vacuum Document Handler (VDH)，克服了現行方式遲緩的問題。

4.1 Top Vacuum Corrugation Feeder(TVCF)

為了提升影印速度，送紙裝置必須要與高速技術相容，不能產生廢紙，而且可傳送所有一般重量的紙張。在本節中，將 TVCF 與速度慢、相容性以及信賴度都較差的送紙方式加以比較，並說明 TVCF 的送紙循環。

4.1.1 TVCF 與其它送紙方式的比較

以往的送紙技術……（以下省略）

【注意事項】

第一段指出以往技術的缺點，強調本公司開發技術的特色。本篇的標題也使用比較級 higher copying speed，顯示與以往技術的比較。與以往技術比較，其實就是引起讀者的興趣，換句話說，就是引發讀者的爭論。第二段在說明送紙與文件傳送技術的條件之後，預告本節內容。讀者在這裡就能了解到這是爭論下一個內容的伏筆，可以作好準備。

這種指出其它方法缺點、突出筆者方法優點的寫法，可以運用於報告、提案以及研究論文等，是非常有效的提問修辭學。此外，比較其它方法與筆者的方法，或是對照式的延伸，叫做 comparison and contrast，這種文章的書寫技巧，能夠有效地說服讀

者。

（參考）何謂簡介

　　簡介 (brochure) 是提供產品或企業相關資訊的小冊子。其實，小冊子這個字最貼切的英文還是 pamphlet；不過在實際的生活中，不妨可以把 pamphlet 和 brochure 當作相同的意義看待。pamphlet 也可以是政治或神學的相關小冊子，而產品的 pamphlet 則是產品的一般說明書。相對地，brochure 則是將重點擺在產品的特色和好處(features and benefits) 上。catalog（美式，英式則為catalogue）雖然也是提供產品資訊的印刷品，但是收錄了許多產品和款式，每一項產品的空間有限。在這有限的空間裡，一般 catalog 只標示照片或簡單的外形圖、規格，偶爾也有價格。brochure 通常是針對單一產品做詳盡的說明。

　　英國人在唸 brochure 時，重音往往擺在最前面，美國人則是把重音擺在後面。而 brochure 大致可以分為產品簡介 (product brochure)、服務簡介 (service brochure) 和公司簡介 (corporate brochure) 三大類，每一樣的標準結構如下：

1. 產品簡介 (product brochure)

　　目的在於希望讀者能對公司產品產生興趣，被產品吸引，進而購買。

　　1) 前言介紹(introduction)──是什麼樣的產品，為什麼希望顧客注意。

　　2) 好處 (benefits)──為什麼要顧客購買，顧客會得到什麼好處。

　　3) 特色 (features)──與其它公司產品的差異。

　　4) 操作原理──依據什麼原理，有哪些功能。

　　5) 適用對象──適合哪些使用者。

6) 適用範圍——有哪些用途。

7) 產品種類——有哪些款式、尺寸和材質。

8) 價格——各種款式的價格，周邊配件的價格。

9) 規格。

10) Q&A——可以想見的問題與回答。

11) 公司概要。

12) 交貨期、裝置方法、保證。

13) 訂購方法。

2. 服務簡介 (service brochure)

依服務的種類不同，內容和結構會有所不同。推銷公司能力的 brochure 叫做 capabilities brochure。

3. 公司簡介 (corporate brochure)

為了讓讀者了解這是一家怎樣的公司，進而推銷公司的能力和形象。

1) 業務內容、行業類別。

2) 公司的組織——母公司、相關企業、公司內部組織。

3) 公司的基本理念 (corporate philosophy)。

4) 公司的經歷。

5) 工廠概況。

6) 用地。

7) 主要市場。

8) 銷售形態。

9) 營業額、利潤與配股。

10) 業界中的地位。

11) 員工人數。

12) 員工福利 (employee benefits)。

13) 發明。

14) 顯著的成績。

15) 研究、開發。

16) 品質管理。

17) 與社區的關係 (community relations)——環保措施、對公眾的貢獻、慈善活動、支援藝術活動。

18) 表揚、資格。

19) 目標及未來的計畫。

以上的說明是參考 Bly (1985) 和長野 (1991)。

2. 引起讀者的興趣

一般公司的外文簡介，大多是忠實地照字面直接翻譯成英文，感覺有些突兀。這無論是委託翻譯的一方或是翻譯社，以及翻譯人員，三方都有責任。

另外，本國人會關心的話題，未必會引起外國人的興趣。無法引起讀者的注意，也就等於無法令讀者閱讀。無法引起他人閱讀興趣的簡介，寧可不寫。

下面的範例，提出了翻譯後簡介的問題點。

【背景】

ABC 機械是抽風機的生產廠商，他們開發控制抽風機啟動時發出極大雜音的技術，也希望將這項技術銷售到抽風機以外的噪音防治領域。將此技術的簡介翻譯為英文，是希望藉此幫助 ABC 公司的國際化。

【原文】

第二次世界大戰後，日本經濟急速發展；但另一方面各種公害問題不斷發生，造成嚴重的社會問題。

之後，由於法規的整備和公害防治技術的發達等，日本名聲

逐漸提高，成為全球最優秀的公害防治國。但是，噪音公害基於
國土狹小、發生源多元化等因素影響，問題複雜，近來尤其受到
重視，申訴案件數目也居第七類型公害之首，達到 40%。

　　ABC 機械是抽風機的專業廠商，長年致力於周邊技術的開
發，而噪音防治也是其中的一環。

　　我們將累積的技術與訣竅集其大成，確立 ANC (ABC Noise
Control System) 系統，提供噪音防治的綜合診斷、靜音器、防音
罩等各種噪音防治對策。尤其在安全閥等氣體排放用以及抽風機
等的抽、排氣用靜音器方面，已經制定標準型，以低價位、短交
貨期的生產體制提供產品，以因應各位的需求。

例文 40　產品簡介（原文）

Rapid development of Japanese economy after World War II
has, on the other hand, brought many serious problems with
"happening" of many kinds of pollution.

After the appearing of such problem, Japan has gained the
reputation as the most advanced country for pollution control by re-adjustment of law and regulation together with the
technical developments for pollution control, however, noise
pollution is recently closed up, because complicated counter-
measures are requested due to the small land of living area,
multiplicate noise sources and etc. According to the recent
data, the number of complaints for noises reaches about 40%
of all pollution complaints which is the top of seven pollution
categories.

ABC Machinery has been studying for long time the noise control action of emission of noise from the blowers, as the one of the circumferential technologies.

ANC (ABC NOISE CONTROL SYSTEM) has been established by our accumulated technologies and summarized know-how. Now, we are ready to supply all kinds of noise control action from all-round diagnoses of noise problem to silencer including noise-cover. Especially, the silencers for gas emission represented by safety valves and for suction and blow out from blower etc. are well prepared as establishment of standard type with low price and short delivery, and the special designed silencers are also presented in compliance with client's needs.

【英文的問題點】

　　無論在文法或語法上，都有很多問題，我們只說主要的幾項。由於 happen 意指「突然發生」，所以第一段的 happening「整備」，在這裡不適用。不要受限於「發生」的字面意義，可轉為 brought many pollution problems 的說法。第二段第一句譯為 re-adjustment of law and regulation，但 re-adjustment 是「再調整」的意思。同一句的 however 加在逗號後面，等於表示與前文的關係，所以應該將前面的逗號改成句號，變成完整的句子，或是改成分號（；），繼續將句子延伸下去。同一句的 closed up 在英文中只有 to close up（及物動詞），表示「關閉」的意思；而 close-up（近照）則是名詞形，都不是「受到注意」的意思。這句話最後的 etc. (et cetera) 的 et 是 and 的意思，所

以前面不需要再加上 and。

【內容的問題點】

這份簡介的結構:

(1)經濟發達, 但是公害越來越嚴重 (背景與問題)。

(2)日本公害防治技術發達, 但噪音公害仍是問題所在 (背景與問題)。

(3)本公司擁有噪音控制技術 (本公司的能力)。

(4)本公司的技術有助於防治噪音 (解決問題的提案)。

(5)尤其標準型價位低廉、交貨期短 (提案的優越性)。

這份簡介的目的在於向受到噪音問題困擾的組織或企業推銷 ABC 公司的技術, 也就是一篇說服性的文章。這份簡介中包含了說服性文章應該具備的所有要素。這些要素包括有待解決的問題、提案者解決問題的能力、問題的解決方案以及其優越性。

根據卡內基 Carnegie (1937) 的說法, 「多數人自我時間的 95% 是用來思考自己的事情」。 (Most people spend 95% of their time thinking about themselves.)人們既然是如此的自我本位, 那麼提案者必須做的, 就是引起讀者的興趣。

接下來介紹改善範例。改善範例的結構為:

(1)噪音公害是全球性的問題, 在日本尤為嚴重 (提出問題)。

(2)噪音防治需要高度技術 (強調問題)。

(3)本公司擁有噪音控制技術 (公司的能力)。

(4)本公司的技術有助於噪音防治 (解決問題的提案)。

(5)尤其是標準型、價位低、交貨期又短 (提案的優越性)。

這個改善範例能引起讀者的興趣, 也就是單刀直入地切入噪音的解決對策。

例文 40 的改善範例 A (△)

Noise pollution[1] is now a crucial social issue the world over[2]. It is so serious that, for example, in Japan, the number of complaints on noises has reached about 40 percent of the total number of complaints received on all kinds of environmental pollution[3]. Furthermore, as the types and sources of noises become more diversified[4], noise control measures increasingly requires advanced technologies[5].

ABC Machinery, a world leader in the manufacture of blowers, has for many years been studying technologies related to the control of noises. Utilizing such accumulated technologies and know-how, ABC has developed the ANC (ABC Noise Control) System, which covers noise diagnosis[6] as well as supply of silencers, noise-proof covers and other noise controlling devices.

Particularly, ABC has standardized[7] silencers for gas discharge devices such as safety valves, and for gas suction/discharge[8] devices such as blowers. Thanks to standardization, ABC can supply these products at low cost with short delivery times.

1) noise pollution 噪音污染 2) the world over＝all over the world
3) environmental pollution 環境污染 4) to become diversified 多元化 5) advanced technologies 尖端技術 6) diagnosis 診斷
7) to standardize 標準化 8) suction/discharge 吸入、吐出

【內容】

　　噪音污染已經成為全球嚴重的社會問題。舉例而言，日本各種關於噪音的申訴已經佔各種類型公害的 40%。此外，噪音形態與噪音源的多元化，使得噪音控制技術也必須更加高度化。

　　本公司是全球性抽風機製造廠，長年以來研究噪音控制的相關技術。我們運用累積的技術與訣竅，開發出 ANC (ABC Noise Control) 系統。它包含了噪音診斷、靜音器、防音罩以及其它噪音控制裝置的提供。

　　尤其本公司已經將安全閥等氣體排放裝置以及抽風機等氣體吸入、排出裝置用靜音器標準化。這將可以使得這些產品以低價位、短交貨期的條件提供給您。

　　下一節介紹另外一個改善範例。

3.　避免強迫讀者接受訊息

　　例文 40 的改善範例 A 重點在於直接從讀者關心的噪音問題下筆。不過，這樣還不一定能引起讀者的興趣，問題就出在第二句的 "in Japan"。因為自己國內噪音申訴的百分比無論有多高，外國讀者根本毫不在乎。

例文 40 的改善範例 B (○)

① Noise pollution has grown from a nuisance[1] to a social problem that must be confronted[2] at all levels of society. In the chemical industry, we at ABC Machinery find our clients increasingly concerned about controlling noise.

② ABC, for many years a world leader in the design and manufacture of blowers and research into their related technologies, has developed the ABC Noise Control System (ANC). ANC incorporates[3] noise diagnosis with a highly effective product line[4] of silencers, noise-proof covers, and other noise-controlling devices.

③ A specialty of ABC is our standardized line of silencers for gas discharge devices such as safety valves, and for gas suction/discharge devices such as blowers. Standardization means ABC can supply these products at low cost and with minimum delivery time, to meet the urgency of your noise control needs.

1) nuisance　麻煩　2) to confront　阻礙　3) to incorporate　內有～
4) product line　產品項目

【段落大綱】

① 噪音污染是社會問題，在化學工業中也頗受注意（提出問題）。
② 本公司開發噪音控制技術（解決問題的提案）。
③ 因為標準化，所以能實現低價格、短交貨期（提案的優越性）。

【內容】

　　噪音污染已經從個人的困擾演變成社會各階層都必須面對的社會問題。在化學工業之中，客戶對於控制噪音也越來越關切。

　　ABC 公司在抽風機的設計、製作以及相關技術的研究方面，長年以來，居於全球的領導地位。此次開發了 ABC 噪音控制系統 (ANC)。ANC 中組裝有靜音器、防音罩、其它噪音控制裝置等高效能產品項目之外，還有噪音診斷。

　　ABC 最擅長安全閥等氣體排放裝置用以及抽風機等抽排氣用靜音器，而且已經標準化，這使得 ABC 公司能夠以低成本和短交貨期來供應這些產品，因應客戶控制噪音的迫切需求。

參考文獻

Bly, R. W. (1985) *The Copywriter's Handbook.* Dodd, Mead & Company.

Carnegie, D. (1937) *How to win friends and influence people.* Simon & Schuster.

Black's Law Dictionary. 5th ed. West Publishing Co.

The Chicago Manual of Style. The University of Chicago Press.

Clavell, J. (1983) *The Art of War.* Delta.

Kirkman, J. (1992) Which English Should We Teach for International Technical Communication? *Journal of Technical Writing and Communication*, No.1–1992.

Longman Dictionary of Business English. [LDBE] Longman.

Longman Dictionary of Contemporary English. [LDCE] Longman.

Military Standard Specification Practices MIL-STD-490A (1985).

Quality Progress. May 1992.

Young, L. E., A. L. Becker & K. L. Pike (1970) *Rhetoric: Discovery and Change.* Harcourt Brace Jovanovichi.

篠田義明 (1977), 《工業英語的語法》, 研究社。

—— (1981), 《技術英語——理論與發展》, 南雲堂。

—— (1994), 《社交英文學習技巧》, 研究社。

篠田義明, J. C. 馬穗斯, D. W. 史蒂芬生 (1986), 《科技英文的技巧》, 南雲堂。

湯姆森・馬提內著, 江川泰一郎譯註 (1989), 《實例英文法》（第 4 版）, 牛津出版社。

長野格 (1991), 《商務英文的語法與語感》, 研究社。

拉提斯, 《單位辭典》。

参考文献

Bly, R. W. (1985). *The Copywriter's Handbook*. Dodd, Mead and Company.

Carnegie, D. (1937). *How to win friends and influence people*. Simon & Schuster.

Black's Law Dictionary (Abridged). West Publishing Co.

The Chicago Manual of Style. The University of Chicago Press.

Chisell, J. (1987). *The Art of Writing*. Delta.

Kirkman, J. (1992). Which English Should We Teach for Interna-tional Technical Communication? *Journal of Technical Writing and Communication*, No.1, 1992.

Longman Dictionary of English Language (LDEL). Longman.

Foreign Division of Coopermany Ltd., (FDCL) Consumin-Mblao, Common Specification Practice (IIT-SLD-001) (1985). (Yunping Press), May 1994.

Young, L. B., A. Becker, K. L., Pike (1970) *Rhetoric, Discovery and Change*. Harcourt, Brace Jovanovich.

王希杰 (1983).《汉语修辞学》. 北京出版社.

——(1991).《修辞学通论》. 南京大学出版社.

——(1996).《修辞学新论》. 北京语言学院出版社.

吕叔湘、朱德熙(1952).《语法修辞讲话》(1980). 中国青年出版社.

陈望道 (1976).《修辞学发凡》. 上海人民出版社.

周振甫主编(1988).《文章例话》. 中国青年出版社.

张弓(1963).《现代汉语修辞学》. 河北人民出版社.

英文 Listening Reading Speaking Writing 零障礙

日文 聽說讀寫 得心應手

國家圖書館出版品預行編目資料

商用英文書信／監修 篠田義明, 高崎榮
一郎 Paul Bissonnette著, 彭士晃
譯.--初版.--臺北市:三民, 民88
　　面;　　　公分
參考書目:面
ISBN 957-14-2966-X (平裝)

1.商業書信　2.英國語言-應用文

493.6　　　　　　　　　　88000331

網際網路位址　http://www.sanmin.com.tw

© 商用英文書信

著作人　高崎榮一郎　Paul Bissonnette
監修　篠田義明
譯者　彭士晃
發行人　劉振強
產著作財權人　三民書局股份有限公司
發行所　三民書局股份有限公司
　　　　地址／臺北市復興北路三八六號
　　　　電話／二五○○六六○○
　　　　郵撥／○○○九九九八——五號
印刷所　三民書局股份有限公司
門市部　復北店／臺北市復興北路三八六號
　　　　重南店／臺北市重慶南路一段六十一號
初版一刷　中華民國八十八年十二月
初版二刷　中華民國八十九年十月
編號　S 80212
基本定價　肆元貳角
行政院新聞局登記證局版臺業字第○二○○號

有著作權·不准侵害

ISBN 957-14-2966-X (平裝)

三民英漢大辭典

林耀福等　主編　定價1500元

蒐羅字彙高達14萬字，片語數亦高達3萬6千。襄括各領域的新詞彙，為一部帶領您邁向廿一世紀的最佳工具書。

三民全球英漢辭典

莊信正、楊榮華　主編　定價1000元

全書詞條超過93,000項。釋義清晰明瞭，針對詞彙內涵作深入解析，是一本能有效提昇英語實力的好辭典。

三民廣解英漢辭典

謝國平　主編　定價1400元

收錄各種專門術語、時事用語達100,000字。例句豐富，並針對易錯文法、語法做深入淺出的解釋，是一部最符合英語學習者需求的辭典。

三民新英漢辭典

何萬順　主編　定價900元

收錄詞目增至67,500項。詳列原義、引申義，讓您確實掌握字義，加強活用能力。新增「搭配」欄，羅列慣用的詞語搭配用法，讓您輕鬆學習道地的英語。

三民新知英漢辭典

宋美瑋、陳長房　主編
定價1000元

收錄中學、大專所需詞彙43,000字，總詞目多達60,000項。用來強調重要字彙多義性的「用法指引」，使讀者充份掌握主要用法及用例。是一本很生活、很實用的英漢辭典，讓您在生動、新穎的解說中快樂學習！

三民袖珍英漢辭典

謝國平、張寶燕 主編
定價280元

收錄詞條高達58,000字。從最新的專業術語、時事用詞到日常生活所需詞彙全數網羅。輕巧便利的口袋型設計,易於隨身攜帶。是一本專為需要經常查閱最新詞彙的您所設計的袖珍辭典。

三民簡明英漢辭典

宋美瑋、陳長房 主編
定價260元

收錄57,000字。口袋型設計,輕巧方便。常用字以＊特別標示,查閱更便捷。並附簡明英美地圖,是出國旅遊的良伴。

三民精解英漢辭典

何萬順 主編 定價500元

收錄詞條25,000字,以一般常用詞彙為主。以圖框針對句法結構、語法加以詳盡解說。全書雙色印刷,輔以豐富的漫畫式插圖,讓您在快樂的氣氛中學習。

謝國平 主編 定價350元

三民皇冠英漢辭典

明顯標示國中生必學的507個單字和最常犯的錯誤,說明詳盡,文字淺顯,是大學教授、中學老師一致肯定、推薦,最適合中學生和英語初學者使用的實用辭典!

莊信正、楊榮華 主編 定價580元

美國日常語辭典

自日常用品、飲食文化、文學、藝術、到常見俚語,本書廣泛收錄美國人生活各層面中經常使用的語彙,以求完整呈現美國真實面貌,讓您不只學好美語,更能進一步瞭解美國社會與文化。是一本能伴您暢遊美國的最佳工具書!

三民英漢辭典系列

Sanmin English-Chinese Dictionary

三民英語學習系列